The
original edition
of this book contained oversized maps.
PDF files of these maps
can be downloaded from
www.elibron.com/maps

ALEXANDR IVANOVITCH
GLOUKHOVSKOY

THE PASSAGE
OF THE WATER OF THE AMU-DARYA

BY ITS OLD BED

INTO THE CASPIAN SEA

AND THE OPENING OF THE DIRECT AMU-DARYA-CASPIAN WATER WAY FROM
THE AFGHAN FRONTIER BY THE AMU-DARYA, THE CASPIAN SEA
THE VOLGA AND THE MARIA SYSTEM TO
ST. PETERSBURGH AND THE BALTIC

Elibron Classics
www.elibron.com

Elibron Classics series.

ISBN 1-4021-5971-4 (paperback)
ISBN 1-4021-0416-2 (hardcover)

This Elibron Classics Replica Edition is an unabridged facsimile
of the edition published in 1893,
St. Petersburgh.

THE PASSAGE
OF THE WATER OF THE AMU-DARYA
BY ITS OLD BED
INTO THE CASPIAN SEA
AND THE OPENING OF THE DIRECT
AMU-DARYA-CASPIAN WATER WAY
from the Afghan frontier
by the Amu-Darya, the Caspian Sea, the Volga and the Maria System
to St. Petersburgh and the Baltic.

An account of the labours of the Imperial Expedition for the exploration of the Old Bed of the River Amu-Darya between the Sea of Aral and the Caspian, in explanation of the transactions of the Expedition, as sent to the World's Columbian Exposition in Chicago, 1893.

Appendices.

1) General map of the Indo-Amu-Darya-Caspian route.
2) Map of the Old Bed of the river Amu-Darya, of the Khivan Khanate and of the Delta of the Amu-Darya.
3) Longitudinal sections of the proposed lines of the passage of the water of the Amu-Darya into the Caspian Sea, through the Sary-Kamysh Depression and by a Canal avoiding the latter

PUBLISHED BY THE CHIEF OF THE EXPEDITION,

Lieutenant-General A. I. GLOUKHOVSKOY.

In English and Russian.

St. Petersburgh

1893

Дозволено цензурою. Спб., 15 Сентября 1893 г.

Типографія и хромолитографія П. П. Сойкина, Стремянная, № 12.

CONTENTS.

PREFACE.

Central Asia, being situated between China, Afghanistan, Persia and Russia, has always in consequence of its geographical position played an extremely important part in the destiny of Asia. A peculiar importance was contributed, as it is now, to Central Asia, by its majestic rivers, the Amu-Darya and the Syr-Darya. Issuing from the snowclad ranges of the Hindu Kush and Thian-Shan they intersect a vast plain, extending for thousands of versts northwards from the foöt of the mountains. A whole range of fertile oases are situated along the banks of these rivers, sharply contrasted with the desert and waterless steppes, which surround them.

The history of the East shows that from time immemorial states flourished in these oases, reaching a high degree of development at a time when in the adjacent deserted steppes wandered half savage pastoral tribes, that have not even to-day issued from their state of ignorance. Thus, if the development of culture and civilization in Central Asia was in direct dependence upon the existence of its magnificent rivers, the reverse phenomenon also occurred not unfrequently.

Both the state of these rivers and the direction taken by them not seldom depended on various historical events, taking place in these countries. Before proceeding to an account of the historical information upon the former course of the Amu-Darya, it is necessary to throw a rapid glance at the political revolutions which took place in Central Asia, and had a considerable influence on the state of the Central Asian rivers and the direction of the course of the Amu-Darya.

Central Asia was not always what it is at present. In its rich oases one of the most ancient civilizations of the world was engendered, and attained a considerable degree of development. Here, in the deepest antiquity, flourished such states as Bactriana, Transoxiana, Sogdiana and Khorazmia or Khovarezm. Among them, Bactriana with its capital Bactra or Bactris, the present miserable Balkh, reached already 1200 B. C., when it formed a part of the Assyrian Empire, a high degree of culture.

Transoxiana in its turn, the present Bokhara, was in a flourishing condition and carried on an extensive commerce. In Sogdiana, in the valley of the Zarevshan, of which Marakanda, the present Samarkand, was the capital, first appeared the mild doctrine of Zoroaster upon the two principles of good and evil, personified in the God of light and the God of darkness. The ancient Khovarezm, situated in the territory of the present Khivan Khanate, into whose composition entered about thirty large cities, was renowned not only for its wealth and trade, but also for its learning. Even the now deserted Transcaspian steppes were not always such. In ancient times, on the shores of the Caspian Sea lay Hyrcania, a country covered with forests full of game. In those times the Amu-Darya, under the name of the Oxus, flowing through that country discharged its waters into the Caspian Sea. Through it lay the trade route between the East and West. Indian goods brought by land to the upper waters of the great river Amu-Darya, were sent by it to the Caspian Sea, whence they came through Transcaucasia to the Euxine or Black Sea. Thus Central Asia kept up in the deepest antiquity close relations with European countries, and served as a connecting link between them and India with all its inexhaustible treasures.

The great migrations of nations took place through Central Asia, and with each migration, with each new wave of people, the states existing on the banks of the Amu-Darya and Syr-Darya fell to pieces and gave place to the organization of new states.

The chief races of Asia, Iran, Turan and Mongol, ruled in succession over the Central Asian oases, came into collision with each other, and swallowed eack other up. Here passed the victorious hordes of Nebuchadnezzar and Semiramis. The conquests of Cyrus the Persian reached as far as the Syr-Darya. According to the testimony of Herodotus, Central Asia attained a high degree of wealth and development. The nations inhabiting it paid to the Persian Kings a tribute of 900 talents (about 2,250,000 metal roubles) i. e. more than Egypt, at that time renowned for its wealth. In the year 334 B. C., as is known, Alexander of Macedon undertook his celebrated expedition into Asia. After having defeated the Persians on the Granicus and at Issus, he passed trough Syria and Palestine and marched against Egypt. Having brought it under his dominion, Alexander again turned his arms against the Persian King Darius, whom he completely defeated in the battle of Arbela. This victory placed Media and Persia in his hands. Alexander then marched in 329 B. C. against Bactriana and Sogdiana, which constituted the last support of the Persian Empire in the far East and, having penetrated with his troops into the valley of the Syr-Daria, extended his dominion over all these regions also. These campaigns rendered inestimable services to civilization, by bringing together for the first time the East and the West, acquainting the ancient Greek civilized world with the nations and countries of the East, about which till then only confused rumours had reached them.

After the death of Alexander, Bactriana became a part of the empire of the Seleucids, but in the middle of the III. Cent. B. C. an independent Graeco-Bactrian state was founded.

The nomadic peoples of Central Asia, the Kirgiz steppes and Mongolia were ever in a state of ferment. They at one time threw themselves upon the rich oases of Central Asia, at another upon fertile China. In the III. Cent. B. C. was raised the Great Wall of China, and the free access of the nomads towards the east into China was cut off. Then, the torrent of wandering nomadic tribes was turned in the opposite direction, into Central Asia. A half savage tribe of Mongolian extraction, in Chinese called the Great Yue-Chje, and known to the Greeks by the name of the Massagetae, under the pressure of another Mongolian tribe, the Giongny or Huns marched at first on the Syr-Darya, and then to the south into Sogdiana and Baktriana, and destroyed the Graeco-Bactrian state at the end of the II. Cent. B. C. Upon its ruins was formed the Indo-Scythian state. The impulse given by these migrations to other wandering half savage tribes called forth subsequently the great migration of peoples to the West, which ended with the fall of the Western Roman Empire.

From the II. Cent. B. C. to the birth of Christ the Chinese, wishing to secure themselves from the constant attacks of the steppe nomads of Central Asia, begin to turn their arms against them and gradually reduce them to their dominion. In the reign of the Chinese Emperor Wu-Di, of the mighty dynasty of Khan, the general Pan-Chan in the I. Cent. A. D. passed victoriously through the whole of Central Asia, and reached the Caspian Sea and the frontiers of the Roman Empire. The rule of the Chinese in Central Asia fell in consequence of the rebellion of the Tiurks or Uigurs, who at the end of the VI. Cent. invaded the valley of the Oxus, and laid the foundation of their dominion there. Henceforth, Central Asia bore the name of Turkestan, or the Country of the Tiurks.

At the beginning of the VIII. Cent. A. D., the Arabs invaded Turkestan, and conquered it towards the end of that century. When the Arabs applied themselves to peaceful occupations, there began an epoch of brilliant development for the civilization of Central Asia. At that time, Khovarezm with its capital Djordjania attained a high degree of wealth and became a centre of Arabian science and culture, like the other centre in Bagdad. The celebrated Arabian astronomers Alkharezmi and Albiruni were natives of Khovarezm. At the court of the Shahs of Khovarezm, other celebrities of the world of science of the time as well as Albiruni found a ready welcome. Among them was Ibn-Sina, an excellent authority on Greek philosophy, better known in the west under the name of Avicenna. The dominion of the Arabs in Central Asia was reflected also on the course of the Amu-Darya. For the better organisation of the irrigation in the neighbourhood of the then capital of Khovarezm, Djordjania, a dam was constructed in the X. Cent. A. D. across one of the chief western arms, the Kunya-Darya, falling into the former Djordjanian Lake or Sarykamysh. The consequence of this dam was a diminution of the quantity of water flowing through, and the interruption of the stream by the Uzboy, which used to fall into the Caspian Sea. Under the dominion of the Arabs, Turkestan formed successively a part of the empire of the Sassanids, Gaznevids and Seldjukids. After the fall of these kingdoms, Khovarezm obtained its independence under the rule of its own Khans. In the reign of the last Khovarezmian Shah, Muhammed-Kutbeddin, Khovarezm was exposed to an invasion of the Mongols. This little known barbarous tribe, inhabiting the depths of Manchuria, gradually subjugating the surrounding tribes, formed a powerful state whose importance in the course of time was ever on the increase.

In the beginning of the XIII. Cent., the Mongolian State under Temu-Chin or Chingis-Khan became a threat to the whole civilized world. The victorious hordes of Chingis-Khan traversed the whole of Central Asia and reached the Black Sea, causing terror to all Europe. The brilliant Arabian civilization fell under the blows of the terrible nomads. After the capture of Bokhara, Chingis-Khan entered its chief mosque on horseback. The city was pillaged and burnt. A like fate befell Samarkand. The inhabitants of Termed, which was taken by storm, were slaughtered to a man. Balkh, which then was in a flourishing state, fearing the same fate, sent a deputation with priests at its head to Chingis-Khan to implore mercy. But the Khan refused, saying: „Your Sultan yet lives and while he is alive the inhabitants of Balkh will always be ready to rise“. After a desperate resistance, Balkh was taken by storm and all its inhabitants put to the sword; the fortress was razed to the ground.

For the conquest of Khovarezm, the victorious Khan sent his three sons, Okeday, Djagatay and Djuchi, who laid siege to the capital Djordjania. According to the statement of Abul-Ghazi, the siege lasted seven months, and all the efforts of the Mongols to take the city were vain in consequence of the quarrels and discord between the princes besieging it. Chingis-Khan then appointed Okeday-Khan commander-in-chief, and the siege was then carried on with more success. When the city was taken, it was set on fire from all sides and the inhabitants led out of it. The artisans, forming 100,000 families, were separated from the rest. The young men and girls were made slaves of. All the rest of the inhabitants were distributed among the Mongolian soldiery, each soldier getting 24 people. All were slain by order of Okeday-Khan. After the downfall of Djordjania, the Mongols destroyed the dam, erected in the time of the Arabs and which cut off the flow of water by the Kunya-Darya, the consequence of which was the restoration of communication between this river and the Caspian Sea. The conquests of the Mongols did not cease with the death of Chingis-Khan. One of his sons, Okeday-Khan, carried the conquest of China to the Yang-Tse-Kyang. Batu or Baty, the son of Djuchi and grandson of Chingis-Khan, undertook an expedition to the west and traversed Russia, Poland and Hungary, leaving behind him everywhere heaps of ruins. Gulaga, another grandson of Chingis-Khan, brother of Khan Mongu, pushed his barbarous hordes into Mesopotamia and Syria, destroyed Bagdad, the centre of Arabian civilization and extended the Mongolian possessions as far as Egypt.

But the monarchy founded by Chingis-Khan also did not last long. In his own lifetime he divided his dominions between his sons. The power of the great Mongolian State soon began to decline to its downfall. In China the dynasty of Ming was established, which threw off the foreign yoke. The dominions of the heirs of Djuchi were divided first into the Golden and Blue Hordes, and then into smaller Khanates. In Central Asia, several independent states were formed, which being at continual enmity with each another, gradually lost all importance. However, at the end of the XIV Cent., a terrible conqueror arose from the dynasty of Djagatay, Timur or Tamerlane, who once more welded together the petty Mongolian states and founded a mighty empire. But, after the death of Tamerlane, this empire again fell into several independent states. In the beginning of the XVI Cent., Turkestan came under the power of the Uzbeks. In the present Kirgiz steppes, a nomadic population was formed of deserters from the various surrounding tribes, which not acknowledging any authority, organized itself into separate bands of freelances after the fashion of the Cossacks. In the middle of the XV Cent., this steppe population became united together and took the name of Uzbeks, or «free men». In the beginning of the following century, the Uzbeks conquered the whole of Turkestan, where they ruled till our own time. Finding themselves in a country with a comparatively high culture, the Uzbeks remained as before nomads and, when settled in towns, continued to live in kibitkas or carts. The Uzbeks had such an extreme contempt for agriculture and peaceful occupations that the name of the natives, Sart, became with them a by-word. The only suitable employment for a free Uzbek was considered to be war and robbery. Besides this the Uzbeks brought with them a principle of heredity, according to which the heir of the Khan was not his son, but his brother. That led to the apanage system, similar to that which formerly prevailed in Russia. The consequence of this was constant quarrels and civil wars between the members of the ruling dynasties. It is evident, that under such conditions the level of culture in Central Asia became necessarily considerably lower. The wealth for which Central Asia used to be celebrated, vanished and invasions on neighbours and the pillage of passing caravans, became one of the means of filling up the treasury. In a word, the once splendid and wealthy Khovarezm, as well as Bokhara and the other Khanates, under the rule of the Uzbeks became converted into nests of robbers. Civil wars, rebellions and robbery became chronic. The Khanates of Central Asia became gradually impoverished, exhausted and finally fell into ruin. At last, the weakness of the Khivan Khans reached such a point, that in 1605 the Yaik Cossacks, 500 or 1000 strong, under the command of their ataman Nechai began a compaign against Khovarezm.

Taking advantage of the absence of the Khan Arab Mahomed and his troops, the Cossacks took Urgench without meeting any resistance. They remained seven days in the town leading a riotous life, after which with a rich booty they started to return, but were overtaken by the Khan, who annihilated the whole band, only two or three men escaping. Some time afterwards, another ataman Shamay with 300 Yaik Cossacks, followed the footsteps of Nechay, but this expedition also ended in complete failure. Shamay himself was taken prisoner by the Kalmyks, and the remainder of the detachment although it reached the Sea of Aral, was overtaken by winter and fell into captivity. «All the cossacks perished there», says the tradition of that campaign. On the whole, the ruinous state of things under the Uzbeks in Central Asia and the restless condition of the weakened Khanates had also an effect upon the course of the Amu-Darya. Wishing to keep the warlike nomadic tribes, inhabiting the lower waters of the ancient beds of this river, in subjection, and not being able to accomplish this by force, the Uzbeks began to construct dams across these channels with the view of depriving them of water. This was one of the chief causes of the interruption of the flow of water by the western arms of the Amu-Darya, in consequence of which the communication between this river and the Caspian Sea, restored during the invasion of the Mongols, was stopped, and the once flourishing neighbourhood of the former capital Kunya-Urgench or Djordjania, was reduced to a desert.

The Emperor Peter the Great was the first, who understood the extraordinary importance of Central Asia, and above all of its magnificent river the Amu-Darya. Having learnt that this river, which fell into the Sea of Aral, might be sent by its old bed into the Caspian, Peter the Great at once perceived the enormous advantages for Russia which must result from such a change in the direction of the river and from the establishment of a direct water way from St. Petersburgh by the Volga and Caspian into the depths of Central Asia. For the realization of this great entreprize Peter the Great organized a complete expedition under the command of Prince Bekovich Cherkassky. In the instructions given to him, Peter the Great ordered him to build unknown to the Khivans a fortress to hold a thousand men at the former mouth of the Amu-Darya, to try to persuade the Khivan

Khan to submit to Russia, to gather various information upon trade and «to examine carefully the course of that» «river (the Amu-Darya), and to investigate whether it» «could be directed into the old bed», into the Caspian Sea. In September 1716 this expedition started on board ship from Astrakhan and in October reached Cape Tiub-Karagan at Mangyshlak, whence lay the great caravan road to Khiva. Here a fort was built dedicated to St. Peter. Having left there a garrison, Prince Cherkassky proceeded further by sea to the gulf called Alexander Bay, where a second fort was founded in a spot abounding in fresh water, and adapted for defence. Again starting on his way, Prince Cherkassky arrived in the beginning of November at Krasny Vody and set about building a third and most important fortress. Leaving in the new Krasnovodsk fortress near the Balkhan Gulf, the former place of discharge of the Amu-Darya, the remaining part of his troops and vessels, Prince Cherkassky started back by land, reached Astrakhan in February of the following year, and occupied himself in preparations for a spring campaïgn.

In Easter week, in the end of April 1717, Prince Cherkassky sent the Cossacks with a baggage train to Guriev, and then proceeded there with the rest of the troops and artillery by sea. When all the preparations for the campaign were finished, the detachment consisting of about 4,000 men marched out on the 11. of June, that is at the time of the greatest heat. The expedition at first marched parallelly to the sea coast, crossed the Emba, came out on the great caravan road from Astrakhan to Khiva, and marching along it reached on the 15. of August the river Karagach in the Khivan Khanate. This march through waterless desert and fodderless steppes was accompanied by incredible difficulties. Nevertheless, notwithstanhing the severest and most frightful conditions under which the march was conducted, the detachment was 66 days on the road and traversed a distance of 1,350 versts in 34 marches, accomplished at the hottest time of the year, when the temperature of the steppes of Central Asia rises above 40° Réaumur. The day after the arrival of the Russian detachment at the river Karagach, the Khivans without any explanation made an attack which lasted two days, but were driven back. The Khan Shirghazi, seeing his losses and the failure of his attacks, had recourse to cunning and a breach of faith. Declaring that the attacks were made without his knowledge, he entered upon negotiations for peace. During the consultations the Khivans repeated their attack on the Russian camp, but nevertheless the negotiations for peace were renewed, and a preliminary treaty was concluded and confirmed by oath on both sides, the representatives of the Khan kissing the Koran and Prince Cherkassky the Cross. The next day, Prince Cherkassky havig received an invitation, started for the Khivan camp accompanied by distinguished nobles present in the detachment, and escorted by 700 dragoons and Cossacks. After the interview of the prince with the Khan, when the preliminary treaty was confirmed again by the Khan himself kissing the Koran, Shirghazi struck his tents and departed with his army to Khiva accompanied by Prince Cherkassky and his little escort. The rest of the Russian detachment was ordered to follow them. Having reached the river Porsungul, near the town of Porsu, the Khivan army encamped. Here the Khan Shirghazi, under pretext of the impossibility of quartering and providing for such a large body of Russian troops, proposed to Prince Cherkassky, to divide his escort as well as the rest of the detachment into several parts. Prince Cherkassky consented and sent orders accordingly to the Russian detachment. Major Frankenberg, appointed to the command in the absence of Prince Cherkassky, was astonished at the order received and disobeyed. Only after the fourth order, in which was included a threat to bring him before a courtmartial, he divided the detachment into four parts and sent them after the Khivans. This served as the signal for the complete massacre of the Russians. Prince Cherkassky had not time to dismount from his horse, after having given the last orders to the Russian detachment leaving the camp, when the Khivans rushed on his small convoy and partly cut them to pieces, partly took them prisoners. Prince Cherkassky with the other distinguished nobles were cut to pieces and their heads exhibited on the market place of Khiva. The same fate met the rest of the Russian corps. They were all attacked by the Khivans, partly cut down and partly taken prisoners and made slaves of.

In the middle of the XVIII Cent., the Shah of Persia Nadir attracted by the former renown of the wealth of the Amu-Daryan Oases, undertook an expedition against Khiva, conquered it almost without a struggle, but soon after abandoned this country, finding no advantage in retaining it. Nadir Shah only devastated Khiva and led away several thousand prisoners to dig canals in Khorassan.

The Nineteenth Century finds a number of independent Khanates in Central Asia, which being sunk in the depths of ignorance and not being able to derive due advantages from the natural wealth of the country, dragged on their miserable existence. In their weakness these Khanates were defended from their mighty neighbours only by their waterless and deserted steppes.

After the death of the Great Emperor Peter I, the

question of the establishment of the Russian dominion on the Amu-Darya and of turning it into the Caspian Sea was abandoned. Only after more than a hundred years Russia, forced by the violence and robbery in the Central Asian steppes, began to move into Central Asia from the direction of Orenburg and Omsk. In the forties, Russia occupied the lower waters of the Syr-Darya, and entered into direct relations with the Central Asian Khanates. From the beginning of the sixties, commenced resolute offensive action from the direction of Western Siberia, namely from Verny. In 1864, af er the taking of Aulie-Ata, Turkestan and Chimkend, the Russian advanced posts in Central Asia on the side of Orenburg and Omsk were united into one line. In 1865 Tashkent was taken, and two years after this fell Samarkand, the capital of Tamerlane.

Having gained a strong hold in Central Asia, Russia saw herself inevitably forced to proceed to the elucidation of the question put forward by the Great Emperor, namely, whether it was possible to restore the former current of the Amu-Darya into the Caspian Sea. After the occupation of Krasnovodsk at the end of 1869, and the taking of Khiva by the Russian troops in 1873, a whole succession of expeditions into the depths of the Turkestan steppes were undertaken to explore the old beds of the Amu-Darya. They were crowned by the sending ot the Expedition organized by order of the late Emperor, which, after five years of operations in the field and the elaboration of all the materials so gathered, finally solved the question of turning the Amu-Darya into the Caspian Sea in the affirmative.

Thus a century and a half elapsed since the time of the Great Emperor, before Russia was again established on the Amu-Darya, and before the experience fo a century proved the extraordinary importance of the Amu-Darya and of joining it to the Caspian Sea.

Having become ruler of the destinies of Central Asia, Russia cannot stop at the theoretical investigation of this question, so important for her and the whole civilized world. Remembering the Will of her Great Reformer, Russia will sooner or later proceed to the realization of this work, which he was not destined to accomplish. With the turning of the waters of the Majestic Amu-Darya through its former course into the Caspian Sea, Russia will not only return to the Amu-Darya Oases their former splendour and importance and develop their natural forces, but by the irrigation of the now sterile Turkoman steppes, will recall to life the whole country, which was in former times in a flourishing condition. The construction of an uninterrupted water way from the Baltic to the confines of Afghanistan will at the same time open to our Fatherland the now inaccessible markets of India, and will attract to this shortest route by the Caspian Sea and Russia a considerable part of the colossal trade of this the richest country of Asia with the European States.

CHAPTER I.

Brief sketch of the historical data upon the former course of the river Amu-Darya and of the explorations of its channels up to the year 1879.

The river Amu-Darya, intersecting Central Asia from south to north and irrigating it, has always been the source of all the life in the Central Asian steppes.

The basin of the Amu-Darya, the existence of which depends upon this magnificent river, was thickly inhabited from time immemorial, and the river Amu-Darya was already known in the most remote antiquity.

The first positive information which has reached us about the Amu-Darya and the direction of its course is found only in the writers of classical antiquity. The river Amu-Darya was known to them under the names of the Oxus and Araxes. These two names being however often confounded with each other, this greatly obscures the information communicated by these authors. But, at any rate, upon the basis of all the information given by ancient authors the conclusion must be drawn that the Amu-Darya in those times flowed in two directions and had its mouths in two different sea basins. In the northern direction, under the name of the Araxes, the Amu-Darya fell into the Aral, and in the western direction, under the name of the Oxus, it poured itself first into the former Sarykamysh Lake, whence it issued as the Uzboy and fell into the Caspian Sea.

The first indications of the course of the Amu-Darya under the name of the Oxus are met with in the «Periegesis» ascribed to Dionysius of Miletus (VCent. B. C.). According to these data the river Oxus after receiving the Araxes falls into the Caspian Sea. Herodotus (mid. V Cent. B. C.) offers more precise information. He speaks of the Amu-Darya under the name of the Araxes as follows: «The river Araxes flows out of the land of the Matieni» «where the river Indus also takes its rise. Cyrus divided it into three» hundred and sixty canals and it issues with forty mouths; all the» «mouths, with the exception of one, are lost in the marsh-» «es and swamps. One only of the arms of the Araxes,» «flows through an open space and falls into the Caspian» «Sea». This quotation shows plainly and categorically, that in those times the Amu-Darya, divided into many arms, fell into the basins of the Aral and Sary-Kamysh, while but one arm, the dried up bed of which forms the present Uzboy, flowed westward through an open and marshy country and poured itself into the Caspian Sea.

The expeditions of Alexander the Great, which acquainted the Greeks with the East, must have considerably widened their knowledge on the Amu-Darya and the adjacent countries. The contemporaries of the great conqueror, Aristobulus, who accompanied him in the Asiatic campaigns; Patroclus, who was in command of the fleet on the Caspian Sea under Seleucus, Polycleitus, and others had the opportunity of themselves visiting those countries, of seeing with their own eyes that which they afterwards described, and accordingly they have left behind in their writings a rich store of material in regard to information about the Central Asia of those times. This material was afterwards elaborated by the celebrated Alexandrian geographer Eratosthenes. Unfortunately, the works of both this author and of his predecessors have not reached us, and we can judge of them only from insignificant extracts contained in the latest authors.

Among these Polybius (in the year 140 B. C.) in his description of the Oxus, which according to his opinion has its issue in the Caspian Sea, mentions some great waterfalls, which form under their impetuously falling waters free passages, through which the nomads together with their horses stole into Hyrcania for making raids, which is most positively confirmed by the survey of the Uzboy or ancient bed of the Amu-Darya, executed by the expedition sent out by Imperial order. In this bed were really found cataracts with the enormous fall of 21 and even 28 feet, as those of Kurtish and Burgun in which occur steep walls across the bed as high as 14 feet; evidently when the water flowed by the Uzboy, running with great speed in the bed it must have left in consequence of its inertia free passages by the walls.

Another author Strabo, one of the most eminent geographers of antiquity, who lived about the year of the birth of Christ and wrote works, which are the principal source for ancient geography, calls the Amu-Darya, as Herodotus had done, the Araxes and speaks of it as follows: It is asserted that the river Araxes, dividing in many places, more than others inundates this country; it falls by all its mouths into the North Sea and only through one into the Hyrcanian Gulf». This passage confirms and completes the information given by Herodotus. In other parts of his work Strabo speaks like other late authors of the Amu-Darya as the Oxus. About the mouth of the Oxus in the Hyrcanian or Caspian Sea he speaks very definitely: «Through Hyrcania flow the rivers Ochus (Murgab) and Oxus till they fall into the sea». But at the same time in another place he declares: «The Yaxartes» «(Syr-Darya) represents from the very beginning a separate river from» «the Oxus, but falls into the same sea», which indicates that the Amu-Darya also fell into the Aral. Thus this apparently contradictory passage amounts to a proof that the river Amu-Darya had in those times outlets in two basins, the Caspian and the Aral.

In Strabo there are allusions, that near to the Caspian Sea existed a sweet water lake so vast, that it was taken for a part of the Caspian. This is a direct reference to the fresh water basin of the Sary-Kamysh. But it is remarkable, that Strabo also relates, from the words of other travellers, the existence in Hyrcania of immense falls, under which whole detachments could pass without getting wet which is confirmed also by the results of the investigations of the above named Imperial expedition. Finally Strabo speaks more than once and with perfect precision of the existence of a large traffic by the ancient Oxus. «Aristobulus and Eratosthenes,» «quoting the words of Patroclus, say that that river» «(the Oxus) is navigable, that quantities of Indian goods» «are conveyed by it into Hyrcania; hence they are car-» «ried into Albania (the present Transcaucasus) and finally» «across the Cyrus (Kura) and the following countries» «into the Euxine (the Black Sea)». From this quotation it may

*

be seen, that the mouth of the Oxus was in the Caspian Sea, and that this river served as a transit route for the trade from India.

Pliny (I Cent. A. D.) mentions in his Natural History, that the river Oxus flowed through a lake of the same name, which refers to the existence of the lake basin of the Sary-Kamysh.

Furthermore in his work there are references to fresh water as existing in the Caspian Sea, which also can only relate to the lake of Sary-Kamysh. Finally this author confirms the information of Strabo, upon the extensive trade in Indian goods, which was carried on in those times, by the Amu-Darya. Pliny says: «It was establish-» «ed in the time of Pompey, that from India Bactriana» «could be reached, as far as the river Icarus which falls» «into the Oxus, in seven days, and that Indian goods,» «transported thence by the Caspian Sea to the river» «Cyrus (Kura) may be conveyed in not more than five days» «to the Phasis (Rion), which falls into the Euxine» (the Black Sea).

Plutarch, who lived in the beginning of the II Cent. A. D., in his biography of Alexander the Great again confirms the preceding data upon the existence of a fresh water basin, which was taken for a gulf of the Sea of Azov, and which must of course be referred to the lake basin of Sary-Kamysh. According to the information of another historian of Alexander, Arrian, a contemporary of Plutarch, the Oxus also falls into «a large Sea near Hyrcania».

According to the data of the celebrated astronomer and geographer of antiquity Ptolemy, who lived in the II Cent. A. D., «the «river Oxus falls into the Hyrcanian Sea» in nearly the same geographical latitude as the Transcaucasian Araxes. This shows, that at least a part of the waters of the Amu-Darya in the time of Ptolemy flowed by its at present dried up bed the Uzboy, into the present Gulf of Balkhan in the Caspian Sea, which is situated nearly in the same latitude as the Araxes. Ptolomy also gives indirect indications of the existence of a certain division of the Oxus into arms, and the situation of this division is stated by him to be in the middle of the river. Finally he also mentions a certain Oxian lake, the centre of which was situated to the right of the general direction of the course of that river and at a short distance from it. It is evident, that under Ptolemy's Oxian lake must be understood the lake basin of the Sary-Kamysh. According to the testimony of Ammianus Marcellinus, who lived in the IVCent. A. D., «the Oxus «falls into the Caspian Sea in Hyrcania». The flowing of one branch of the Amu-Darya (the Oxus) into the Caspian and of the other, the Araxes, into a separate sea basin finds new confirmation in the data of the Carta Peutingeriana of the end of the IV Cent. A. D. This map shows, that the Oxus falls into the Hyrcanian Sea and that its mouth is situated opposite to the issue of the Cyrus (Kura) in the Transcaucasus. To the east of it is shown the Araxes falling into the Eastern Sea.

From the review above made of the information on the Amu-Darya furnished by ancient authors, it is impossible not to come to the conclusion, that for the time, for which data exist on the course of this river, branches of the Amu-Darya may be assumed. All authors, Roman as well as Greek, confirm this fact in one form or another. The Amu-Darya possessed in those times two separate mouths in two different sea basins. The northern arm, which flowed approximately by the present bed, fell into the basin of the Aral; the western branches flowing by its old bed, fell into the basin of the Sary-Kamysh, whence the water flowed by the Uzboy, and in its lower part forming immense waterfalls, fell into the Caspian Sea.

As Rœsler justly remarks, we find the best proof, that the testimony of the ancient authors that the mouth of the Oxus was in the Caspian Sea should not be rejected, in the circumstance, that the southern part of the Caspian basin, but only the southern part, was explored by the ancients and that they might be acquainted with the mouths of the Oxus not merely from hearsay. A proof that they really did navigate the southern part of the Caspian Sea, is their statement of the length of its shores from the outlet of the Mardus (Zefid-rud) to the mouth of the Oxus, which very nearly agrees with the actual distance between the mouth of the Zefid-rud and the gulf of Balkhan, i. e. the former outlet of the Amu-Darya. A second proof is the latitude, in which Ptolemy places the mouth of the Oxus, which is nearly the same as the latitude of the mouth of the Transcaucasian Araxes and that, in its turn, leads to the gulf of Balkhan and to no other place on the eastern shore of the Caspian Sea. A third proof is the definition of the latitude of Talka, an island in the Caspian Sea, nearly corresponding to the latitude of the mouth of the Oxus. According to the description, the island Talka is evidently the present Cheleken, situated to the southwest from the entrance into the bay of Balkhan. All this, taken together, must lead to the final conviction that the mouth of the ancient Oxus in the Hyrcanian Sea was really the mouth of the present Uzboy.

After the authors of classical antiquity, mentioned above, information upon the direction of the course of the river Amu-Darya ceases completely till the time of the Arabians, when data upon this river are again to be met with in the works of Oriental writers. As regards the direction of the course of the Amu-Darya, the most ancient of the Arabian authors of the end of the IX Cent. A. D. Ibn-Khordadbeh and Yakubi state, although very shortly yet very precisely, that this river falls into the Caspian Sea. The statement of these authors finds confirmation in the circumstance, that during the exploration by the Imperial expedition on the Uzboy, between the wells of Kurtysh and Bala-Ishem, in a cemetery alongside the ruins of Talai-Khan-Ata a Khivan mullah found on a gravestone the year 170 of the Hegira, which corresponds to the middle of the VIII Cent. A. D. In this desert the only possible source of water was the Uzboy and therefore it must be admitted, that at that time water flowed over this bed.

Based on the sum of these data, which all indicate the existence of a communication between the Amu-Darya and the Caspian Sea in the VIII and IX Cent. it may be assumed, that this communication did not cease during the interval of time, for which we have no information of this river. Thus it may be assumed, that one of the branches of the Amu-Darya had its mouth in the Caspian Sea during all the time from the V Cent. B. C. to the X Cent. A. D., i. e. during fourteen centuries without interruption. But since the X Cent., circumstances evidently changed, for not one of the writers of the East refers to the mouth of the Amu-Darya or of its branch into the Caspian Sea.

The Arabian geographers from the X century to the XIII, Istakhri, Massudi, Ibn-Haukal, Ibn-Dosteh, Edrissi and Jakut declare, that the Jyhun, i. e. the Amu-Darya, has its mouth in the lake of Khovarezm, which they describe as follows: It had in circumference about one hundred parasangs (about 528 miles); the water in it was bitter and had no visible outlet; into this lake, besides the Jyhun

(Amu-Darya), fell the river Shash (Syr-Darya), and the distance between their mouths was about ten days' journey. The northern shores of the lake were inhabitated by Tiurk nomad tribes, the Ghuzzes; the distance between this lake and the Caspian Sea, or as it was called by the Arabs, the Khazar Sea, was. in a straight line twenty days' journey.

From this description as well as from the distances stated, which approach very nearly to the corresponding distances of the present time, it must be concluded, that the lake of Khovarezm, above mentioned, is no other than the present Aral. The writers know nothing about any branch of the Amu-Darya falling into the Caspian or Khazar Sea; and in their description of the eastern shore of that sea they represent it either as totally desert or as inhabited by nomads.

The eastern geographers have left us a sufficiently detailed description of Khovarezm of that time, upon the basis of which we may reconstruct a more or less exact sketch of the canals and branches of the Amu-Darya in this country. Khovarezm was situated on both banks of the Jyhun (Amu-Darya). Its dominions up the river reached the town of Tahenia four days' journey from Hazarasp, from which began the cultivated strip on the left bank of the river. On the right bank the inhabited places began a little higher up from the present Tiui-Muyun, known under the name of Abukashi. Below this pass the canal of Gou-Khavare branched from the right bank of the Amu-Darya, from which canal water was taken for the irrigation or the country up to the very capital. This canal, directed nearly parallelly to the former bed of the Amu-Darya, reached the present range of Sheikh-Djeili and its water began apparently to ooze through the hills and to form subterranean passages. Edrissi at any rate, whose works were written later than those of other Arabian geographers, mentions a river Bura, which flows out of a mountain and falls into the Jyhun close by to Djordjania; this river is not mentioned at all by other authors. The ancient capital of the whole country Kiat, otherwise called Khovarezm, was situated on the right bank of the Amu-Darya near to the present town of Sheikh-Abbas-Ali. The ancient castle was undermined by the river, and finally quite destroyed. The large and populous town suffered also exceedingly from the same cause. To the west of the Jeyhun branched the following canals and arms.

1) The canal of Hazarasp, which passed close by the town of the same name.

2) A little below was the canal Kerderan-Khash, larger than the preceding.

3) Then the Khiva canal, which was conducted to Khiva and was larger than the Kerderan-Khash.

4) The arm of Medra larger than the preceding one, on the bank of which was situated the flourishing and populous town of the same name[1]).

5) A little below it, parted the arm Wadak, by which ships descended to the second capital of the country, Djordjania. In the neighbourhood of this town is an arm or a canal named Buh, which joined the Wadak at a distance of one day's journey from the capital. The Wadak was evidently situated in the direction of the present dried up bed of the Kunya-Darya, because upon the site of Djordjania, after its destruction by the Mongols, was founded the town of Kunya-Urgench; and the canal of Buh evidently flowed in a direction nearly corresponding to that of the present canal Lauzan. Djordjania, the second capital of the country, was composed of two parts, situated on both sides of the river: the part of the town on the left side was called Djordjania and that on the right Dargah; communication between them was maintained by means of boats. The town was large and rich and served as a centre for the trade with the nomad tribes of the Ghozzes.

6) Somewhat lower than the Wadak, the canal of Kerder, parted from the Jyhun in four little canals, which soon joined into one common canal. After this the Kerder made a turn, crossed the Wadak arm (the Kunya-Darya) and passed close by to the towns of Little Djordjania (Gurganjek, Kurganchik) and Mademinia (Medemin-Kala). The Kerder canal thus passed nearly in the direction of the canals Mangyt-Arna, Shamrat and then passed into the lower part of the bed of the Daudan. The cultivated lands ended at the village ot Hit in the neighbourhood of the present mountain of Tuz-Gyr-Kry. Of the remaining towns of the Khovarezm of that time deserve to be mentioned Mezdakhan, between the Kerder and the Jyhun, and the town

---

[1]) Judging from the fact that this arm or canal was larger than those preceding, it must be supposed that it passed by the upper part of the presant dry bed of the Daudan.

of Kariat-Farategin on the southern slope of Ust-Urt, in the place of the present ruins of Deu-Kesken; these towns served as centres for the trade with the nomad tribes of the Ghozzes.

From this description of the canals and arms of the Amu-Darya it may be seen that at that time water flowed over the at present dried up beds of the Kunya-Darya and the Daudan. These beds had the aspect of rivers, over which navigation took place. The quantity of water, flowing by the Kunya-Darya was so considerable, that writers frequently mistake this bed for that of the Jyhun, (Amu-Darya).

As the whole locality occupied by these beds has a common slope to the Sary-Kamysh and not to the Aral we are obliged to admit that in the above mentioned beds the water flowed down precisely into this hollow and should have formed a considerable lake, although separated from the Aral Sea by an insignificant isthmus, yet at any rate forming an indepedent basin. In consequence of the absence of an outlet from this lake by the Uzboy into the Caspian, the water in it peshaps was not quite fresh but brackish. And yet it must have been there, and some indications of its existence can be found in the writers of that time. This lake had even its separate name. The Aral, into which the principal channel of the Amu-Darya fell, which passed by the capital Kiat, called otherwise Khovarezm, bore the name of Lake of Khovarezm or Kharezm. The lake basin of Sary-Kamysh, into which fell the Kunya-Darya, was named after the other capital, Djordjania, the Lake of Djordjania, as appears from some passages of the works of Messudi and Edrissi.

On the basis of all that is above set forth, it is impossible not to come to the conviction, that, if from the X. to the XIII. century there was no stream of water in the Uzboy, the lake basin of Sary-Kamysh, fed by the well filled arms of the Daudan and Kunya-Darya, must have existed, and the level of the water in it, if it was lower than the Aral, was in any case not so far lower as to deserve to be reckoned as dried up, as at the present moment. As to the reasons for the stopping of the stream in the Uzboy and the disappearing of the communication between the Amu-Darya and the Caspian, which was of such great importance not only for Central Asia, but also for the whole civilized world, some of the above mentioned authors give us some very positive indications upon the matter.

They mention namely a dam, barring the current in one of the chief arms falling into the basin of the Sary-Kamysh.

Istakhri, describing the Wadak, makes the following statement about the dam: «Boats can approach (by the Wadak) the town» «(Djordjania) only as near as the flight of an arrow, because a dam» «is erected there». And Mukaddasi in his description of the same town says in greater detail about the dam: «Djordjania, the capital» «of the Khorassan side, (the left bank) is situated on the» «Jyhun (the Kunya-Darya), so that the water previously came» «up to the walls. But the current of water was arrested» «by the construction of an artificial dam out of brushwood» «and timber and led away eastwards; this work is indeed» «worthy of admiration. In consequence of this the water was led» «away through the desert to Kariat-Farategin, so that it flows now» «only on one side of the town. Then canals were conducted to the» «town, which pass now near to the gates but do not pass through the» «town itself, because it is too thickly built»

Thus above Djordjania in the present bed of the Kunya-Darya a dam was erected, which arrested the water in it. The current was led away to the right to Kariat-Farategin, which was situated on the southern slope of the chink of Ust-Urt on the site of the present ruins of Deu-Kesken, and consequently the water from this canal finally fell again into the Kunya-Darya. It should be observed at the same time, that Mukaddasi calls this building «worthy of admiration». This seems to show, that it was unique of its kind, that is, the first dam in Khovarezm erected on one of the chief arms of the Amu-Darya [1]).

The question now is: Could such a comparatively insignificant cause, as the erection of a dam in one of the arms of the river, have called forth a phenomenon of such a vast, worldwide importance, as is the interruption of the communication of the Amu-Darya with the Caspian Sea?

Under the circumstances, which are generally met with in river basins it indeed seems to be improbable, but in the exceptional climatic and topographical conditions, in which that branch of the Amu-Darya, which fell into the Caspian Sea, must have been, it may be

---

[1]) It is very possible that similar dams were constructed in other smaller arms and canals.

fully admitted. In reality the Amu-Darya flows through the centre of the vast Asiatic continent, through waterless steppes, under the burning rays of a southern sun; all these circumstances cause the exceeding dryness of the Central Asian climate, under which the expenditure on evaporation in the river must reach exceedingly great dimensions. At the same time the western arms of the Amu-Darya met with topographical conditions, which were exceptionally favourable to this evaporation. The great basin of the Sary-Kamysh was situated on the way to the Caspian and was the cause of the formation of a great lake basin, in which the expenditure of water in evaporation was bound to take such considerable dimensions, that it could easily swallow up the greater part of the total quantity of water flowing by the western arms of the river. Under these conditions a reason causing even an absolutely insignificant diminution of the expenditure of water must have directly and exclusively affected the quantity of water flowing by the Uzboy out of the lake of Sary-Kamysh, because the evaporation in the lake must have remained without change. It might at the same time have easily happened, that the diminution of the expenditure in the arms, which supplied the lake with water, proved to be so considerable, that it swallowed up the whole expenditure of the Uzboy. In this case that channel must have dried up the water communication between the Sary-Kamysh and the Caspian Sea. And the whole expenditure of water of the western arms of the Amu-Darya was bound to go in evaporation in the lake basin.

Let us illustrate these arguments by a numerical example. Let us suppose that by the Kunya-Darya, before the erection in it of the dam, passed on an average yearly 40 cubic sagens of water per second and that the yearly expenditure of water in the Daudan was equal to 36 cubic sagens per second.

The total quantity of water passing into the Sary-Kamysh will be:

By the Wadak (Kunya-Darya)	40	cub. sag.
» » Kerder (Daudan). . .	36	» »
Total on an average for the year .	76	cub. sag. per second.

The surface of the Lake Sary-Kamysh of that time, according to the investigations of the Expedition sent out by order of HIS

MAJESTY THE EMPEROR OF RUSSIA, might have been, with a level of $10^1/_2$ feet above the level of the Aral Sea, equal to 14,000 or so square versts, or about 3,500,000,000 square sagens. The evaporating layer of water in the Oasis of Khiva, according to the observations of Dorandt, amounts to about 0.6 sagen a year. But as, from the considerable dimensions of the former lake of Sary-Kamysh the lower layers of the atmosphere, adjacent to the surface of the water, must have been to a certain extent saturated with aqueous vapours, it may be assumed with considerable probability, that the evaporating layer in that lake was not greater than 0.5 sag. In that case, in this lake, there will go in evaporation yearly $3,500,000,000 \times 0.5 = 1,750,000,000$ cubic sagens of water, which makes on an average $\frac{1,750,000,000}{365 \times 24 \times 60 \times 60} = 54$ cubic sagens per second. Besides this, water must have been expended upon the irrigation of the neighbouring fields, in evaporation in the canals, and finally in filtration through the lake and the canals. Let us suppose, that the expenditure for all this amounts to on an average yearly 13 cubic sagens per second. Then, the total expenditure must have been:

Evaporation in the lake . . .	54 cub. sag.
Irrigation, evaporation in the canals and filtration	13 » »
Total on an average for the year .	67 cub. sag. per sec.

Consequently, under such circumstances, a surplus of water is obtained equal to 76—67 cubic sagens, which must have found its way by the Uzboy into the Caspian Sea. Supposing now that, with the erection of the dam near the town of Djordjania, the expenditure of water in the Kunya-Darya diminished from 40 to 30 cubic sagens per second, the quantity of water directed to the Sary-Kamysh would be as follows:

By the Wadak (Kunya-Darya) . .	30 cub. sag.
» » Kerder (Daudan)	36 » »
Total on an average for the year .	66 cub. sag.

per second, which was not able to cover the expenditure on evaporation, filtration and irrigation. In consequence of this, the stream of water it the Uzboy was bound to stop, and the basin of the lake Sary-Kamysh diminish in its volume until the expenditure of water on evaporation from its diminished surface corresponded no longer to the quantity of water flowing to it. Thus it results, that the erection of a dam in the bed of one of the arms bringing water into the lake of Sary-Kamysh and a comparatively insignificant diminution of the quantity of water flowing through it might in fact cause an interruption of the stream of water by the Uzboy and the destruction of the communication of the Amu-Darya with the Caspian.

The question remains to be solved, What could have caused the necessity for the erection of such a dam? The natives explained to the above mentioned Imperial Expedition the erection of dams in the ancient beds and canals in two ways, either political considerations, for securing themselves from rapacious nomads and robber tribes inhabiting the lower parts of the river, or in consequence of the economical needs of irrigation for the artificial raising of the water in the channel by which, in consequence of the diminution of its quantity, the water no longer flowed with a full current. The first of these explanations cannot, strictly speaking, be applied to the dam erected above Djordjania, because its erection coincides with the period of the prosperity and power of Khovarezm, when restless nomad neighbours could not have been dangerous.

The second explanation remains, which we shall examine more minutely. As is well known, the system of irrigation at present generally practised in Central Asia is very imperfect and it may be supposed that it was in the same stage of development in the X Cent. One of the chief defects of the present system is in the complete absence of any arrangement for the regulation of the supply of water going to irrigate the fields. The aryks or little canals, branching off from the river or chief irrigating canal, are open and are kept so during the whole time that water is needed for irrigation. To remove the deposits that have collected and to repair the canals, they are dammed up late in the autumn. If consequently the Khivan is unable to direct the supply of water and is in this respect in dependence on the natural rising or falling of the level of the river, he can regulate the quantity of water indispensable for irrigation, which changes according to the nature of the soil, the season of the year and the crop, only by means of regulating the expenditure of water already used or unnecessary. By arbitrarily augmenting or diminishing this expenditure, the Khivan can direct to irrigation the exact quantity needed by local conditions. In one word, an essential condition of good irrigation under the present system, is the facility of letting off somewhere the used and superfluous water. The general character of the locality in the Khivan Oasis is such that the most elevated spots occur near the river and, retiring from the bank, the ground gradually falls, but nevertheless its general slope is not sufficiently great to enable the superfluous or used water to run off quick enough. The Khivan accordingly always tries to dispose his fields near some depression, or old channel, where he may let off with ease and convenience the whole quantity of useless water. Thus if there are no hollows whatever in the vicinity, for the leading off of superfluous water, the arrangement of good irrigation is necessarily difficult. The difficulties increase if, for some reason or another, the quantity of water flowing by the arm or canal diminishes. The dimensions of the channel corresponding to the former greater quantity of water, the latter is no longer able to fill up the bed, and therefore its level must fall. In such conditions, it is no longer possible to lead open canals from the channel, but it is necessary to arrange devices for lifting the water, e. g. water pumps, to raise the water above the level of the fields irrigated, involving a considerable and constant expenditure of labour and material means which, in case of raising the water above a certain limit, may even render irrigation perfectly unprofitable and consequently impossible. But the circumstances completely change, if the river is dammed up with a dike and the water carried away by an artificial canal. Then, the whole locality on the river side near the dam will suddenly be placed in exceptionally favourable conditions for irrigation. Out of the new artificially dug canal the water, raised to the desired height, will flow off by open side canals without any water-raising contrivances to be employed in irrigation. The superfluous and used water can easily be let into the channel of the river below the dam, where in consequence of the erection of that construction the level must have fallen considerably.

Thus, the erection of the dam at the town of Djordjania, it must be supposed, was called forth by the demands of irrigation according to the imperfect system which is practised even now in the Khivan Khanate. The consequence of the erection of this dam was, on the one hand, the economical dawn of Djordjania and its neighbourhood and, on the other, the stopping of the stream in the Uzboy, and the transformation of flourishing places along its channel into a desert.

In the beginning of the XIII Cent., Central Asia fell under the yoke of the Mongols. The victorious hordes of Chingis-Khan traversed the whole of Central Asia, and the brilliant Arabian civilization fell under the blows of the terrible nomads.

In 1220 the Mongols besieged Djordjania, took it after a prolonged siege, killed or enslaved the inhabitants, and levelled the town with the ground.

According to Ibnu-l-Atir and Dimashki, the Mongols, after having taken Djordjania and slaughtered the inhabitants, destroyed the dam which was built above the town and held up the water. The result of this was that the whole neighbourhood of Djordjania was inundated, the buildings fell down, and on the site of the city was formed a marsh. All who escaped the Mongol sabres perished in the waves or under the fallen walls.

What subsequently became of the dam destroyed by the Mongols is unknown. At any rate there is no direct evidence. But, judging by the circumstance that several writers and oral traditions connect the change of the course of the Amu-Darya with the invasion of the Mongols, it must be supposed that the dam was not renewed.

In the work of Yakut who, as above mentioned, so definitely affirms the falling of the Amu-Darya only into the Aral Sea, we find in the description of Mangyshlak the first evidence that the Amu-Darya began to fall into the Caspian Sea. This passage is of course in contradiction with what he wrote in other parts of his work. But this notice is important for us, as a proof that, among the materials which Yakut made use of for his extensive compilation, were some data testifying to the falling of the Amu-Darya into the Caspian Sea.

Hamdallah-Mostawfi-Kazwini, who wrote a hundred years after the Mongolian invasion, was the first to categorically state that one arm of the Amu-Darya fell into the Caspian Sea. At the same time he affirms that this happened at the time of the foundation of the Mongolian state. At any rate, since the time of the Mongolian invasion, appears credible information that the communication of the Amu-Darya with the Caspian Sea interrupted in the X Cent. was afterwards reestablished.

The first positive information of the Amu-Darya falling into the Caspian Sea occurs in Mahmud-al-Shirazi, who died in 1310. In his astronomical work, written in the Persian language, he speaks of the successive falling of the Amu-Darya into the Caspian and the Aral Sea. Certain indications of the communication of the Amu-Darya with the Caspian are also met with in Dimashki, who published his Cosmography about the year 1325. He himself mentions only one mouth of the Amu-Darya, in the Aral Sea, but at the same time he states that he had also heard of its western arm falling into the Caspian Sea.

But the truest and most exact information on the course of the Amu-Darya occurs in the Persian geographer Hamdallah-Mostawfi-Kazwini, whom Jobert justly calls the Persian Eratosthenes. Regarding the mouths of the river Amu-Darya, that author makes the following statement: «Some of the arms of the Jyhun (Amu-Darya)» «flow into the Khovarezmian lake (the Aral), but the chief channel» «of the river Jyhun, after leaving Kharezm, forms a waterfall at» «Khulum, which is called by the Turks Gerlaveh; its roar is heard» «from a distance of two or three parasangs. After that it falls into» «the sea of Khazar (the Caspian Sea) at the place called Khul-Khan;» «this place is inhabited by fishermen»! In this quotation from the Persian geographer we find, on the one hand, the categorical statement that the mouth of one of the arms of the Amu-Darya was in the Caspian Sea, and on the other, a new confirmation of the fact, that this arm flowed by the at present dried up bed of the Uzboy, as appears from the references to the presence of great waterfalls, the existence of which in the Uzboy is proved by the investigations of the Imperial Expedition. We find a confirmation of the statement of Hamdallah, of the falling of the Amu-Darya into the Caspian Sea, in his description of the Caspian Sea: «Great rivers fall into this «(Caspian) sea namely, the Atil (Volga), Jyhun (Amu-Darya), Kura,» «Araxes, Shakh-rud, Zefid-rud and others». Speaking of the Lake of Kharezm, Hamdallah expresses himself very precisely, to the effect that into this lake falls only a part of the waters of the Amu-Darya,

«The lake of Kharezm. This lake has about a hundred para-» «sangs in circumference. A part of the waters of the Jyhun,» «the Shark (Syr-Darya), the river Fergana and other tributaries, dis-» «charge their waters into it. The water of these rivers is fresh and» «agreable to drink, while the water of the lake is very salt. The shores» «of this lake are separated from the shores of the Caspian Sea by an» «isthmus, the breadth of which is about one hundred parasangs». Thus Hamdallah speaks very positively about two arms of the Amu-Darya, of which one fell into the Caspian Sea and the other into the Aral.

Besides this, in Hamdallah's description of the Amu - Darya occurs interesting evidence, that this river, after leaving the defile of the «Lion's Jaw», the present Tiuya-Muyun, at some distance from it sinks into the ground. This information, it must be remarked, is confirmed by the late authors, Hadji-Khalifa and Jenkinson. Comparing this information with what Edrissi mentions, when he speaks of the formation of an entire river, the Bura, out of springs situated along the mountain, the conclusion is inevitable that in the XIV Cent. the water succeeded in wearing in the lime rocks of the Sheikh-Djeili range such extensive passages, that the chief stream of the river found its course through them. As a supplement to all the above mentioned information, Hamdallah, in his description of the Caspian Sea, declares that in consequence of the falling into the Caspian Sea of one arm of the Amu-Darya, during the preceding century, the level of this sea in his time was raised to such a height, that the celebrated port in Abbaskun and the neighbouring country were submerged. Let us observe on this occasion, that some of the facts supplied by Lenz Senior, in his investigations of the changes of level of the Caspian Sea, confirm remarkably in this respect the statement of the Eastern author. Thus among other things, Lenz found in the southern part of the bay of Baku formations of limestone with inclined strata, containing in their cracks and hollows horizontal streaks of clay, containing shells such as still exist in the Caspian Sea. These deposits of clay are direct evidence of their marine origin, and that when they were formed the level of the sea was about twelve feet higher than at present (written in 1830). The same is indicated by the city wall situated along the sea shore, by the colour of its lower part, namely that it too was once washed by the sea waves, although it now stands on the dry shore. Indications relating to the time, when the level of the sea was at such a height, are met with, as Lenz observes, in the Arabian geographer of the XV Cent., Bakui. These indications confirm in a remarkable manner the statements of Hamdallah. Bakui says, that in the year 1400 the level of the Caspian Sea was raised to such a height, that part of the town of Baku was submerged, and that in his time the water reached the mosque, i. e. it would seem, was in the streets of the present town. This height, according to the measurements of Lenz Senior, form about 15 feet above the level of the Caspian Sea in 1830.

Information of the falling of the Amu-Darya or of one of its arms into the Caspian Sea, is met with also in western European sources. The first western European document, supplying information on the communication of the Amu-Darya with the Caspian Sea, is the map of the Venetian *Marino Sanudo,* in the year 1325. On this map, the Hyrcanian or Caspian Sea is represented in a place corresponding with the real situation of that sea. To the east of it is situated another Caspian Sea, a lake covered with islands, approached from the west by the Caspian Mountains. Out of this lake flows a river, passing through these mountains to the west, and falling into the great Hyrcanian or Caspian Sea. Still farther to the east of the lake covered with islands, is to be seen a third small nameless lake, into which the Jyhun falls. On this map the nameless lake is evidently the Aral Sea, while the second Caspian Sea, out of which flows the river, which falls into the real Caspian, is nothing else than the independent basin of the Lake Sary-Kamysh.

The Italian and Spanish portulans, or the descriptions and maps of the sea-shores and ports, compiled from the intelligence communicated by navigators and merchants, namely: Portulano Mediceo (1351), Pizzigani (1375) and the Catalonian (1375), and finally the Venetian portulan composd by Fra Mauro (in 1459) show that the western nations were well acquainted with the river of Urgench, that is with the Amu-Darya, and knew it as an affluent of the Sea of Baku, that is of the Caspian.

In the beginning of the XV Cent., the king of Spain sent as ambassador to Timur in Samarkand, Ruy Gonzales de Clavijo, who in the diary of his travels mentions that the river Amu-Darya, which he calls Viadme and Biamo, and which he crossed, falls into the Caspian Sea.

Very remarkable information on the course of the Amu-Darya and Syr-Darya are furnished by a Persian manuscript, belonging to Rawlinson. Its unknown author, in his description of Asiatic lakes, states about the lake of Khorazm, that is the Aral Sea, as follows: «It is said in all ancient books, that the Jyhun» «falls into the lake of Khorazm, but at the present time» «(in the year 820 of the Hegira, or A. D. 1417), this lake no» «longer exists, because the Jyhun has made itself a new» «channel to the Caspian Sea». Besides changes in the course of the Amu-Darya, this manuscript points out a no less change in the course of the Syr-Darya. «The river Khodjend in the» «lower parts of its course, crossing the desert of Kharezm,» «joins the Jyhun, and thus falls finally into the Caspian Sea». However unexpected the information supplied by this manuscript, and however contradictory of the present disposition of the waters in Central Asia, there are yet facts, which confirm them to a certain degree. Thus Mahmud-al-Shirazi says about the Syr-Darya the same as about the Amu-Darya, i. e. that it falls at one time into the Caspian, and at another into the Aral Sea. The Arabian writer of the end of the XIII Cent. Shumsuddin-Dimashkhi testifies also, that the Syr-Darya joined the Amu-Darya.

The sultan Baber, the founder of the Mongolian empire in India and an excellent authority on Central Asia, declares that the river Syr-Darya is lost in sandy deserts. The ancient Chinese maps represent the Syr-Darya and Amu-Darya as two rivers with parallel courses, but the Syr-Darya receives an arm of the Amu-Darya before falling into the Western Sea. Besides which, European sources indicate that just in those times the Aral Sea, if it existed, at any rate was of such inconsiderable dimensions, as to totally escape the attention of various travellers. At that time, travels from Europe into the depths of Central Asia were not so very rare, and many of the travellers left memoirs; but not one of them mentions the Aral Sea, although the road of most of the travellers lay near it or through it. The maps of those times supply the same information as the travellers. The Aral Sea is not represented on one of them. It can scarcely be admitted that such an immense mass of water, as the present Aral, could pass unnoticed, if it really existed, or be held insignificant by these writers.

As to the question of the possibility of communication between the Syr-Darya and the Amu-Darya, it should evidently be answered affirmatively. There exists an old channel along the southern slope of the Sheikh-Djeili range, and joining the Amu-Darya opposite the exit of the canal Mangyt-Arna, that is the so called Akcha-Darya. It is possible that it formed the mouth of the ancient channel of the Syr-Darya, which once sent one of its arms westward to join the Amu-Darya. However this may have been, the quantity of water, which flowed in those times by the western arms of the Amu-Darya through Sary-Kamysh into the Caspian, increased in reality, in a comparatively short interval of time, to such an extent, that the level of water in that sea reached at that time, according to the researches of Lenz, its maximum. Such a great flow of water into this sea, causing an extensive change of level, could only proceed from the fact, that a comparatively very large volume of water was sent by the present Uzboy. But in such a case, the flow of water into the Aral by other arms of the Amu-Darya, must necessarily have decreased by the same amount of water which, in consequence of its smaller surface, must have manifested itself by still abrupter changes in the outline of this basin.

Thus the rapid drying up of the basin of the Aral, which is confirmed from other sources, as well as the simultaneous rise of level in the Caspian Sea, were nothing else but a natural consequence of the increased flow by the Uzboy, and the conjunction of these, at first sight, improbable circumstances, serves as a proof of the existence at that time of a communication between the Amu-Darya and the Caspian.

To the end of the XVI Cent. belongs also the Russian information upon the Amu-Darya and the direction of its course. This is contained in the «Book of the Great Map», a commentary on the map of the State of Moscow with the neighbouring countries; the map itself has unfortunately not come down to us. Although the Book of the Great Map was compiled in the beginning of the XVII Cent., the information contained in it may be well referred to the XVI Cent., when the materials for it were gathered. This is what is said in the Book of the Great Map about the Amu-Darya. «Out of the» «Blue Sea (Aral) flowed the river Arzas (Argas) and ran» «into the Sea of Khvalim». The name Arzas or Argas nearly approaches in pronunciation to the Araxes of the ancient writers and the Ardok of Jenkinson, which shews directly that all these names relate to the same river. In the given case this name refers only to the western arm of the Amu-Darya, which fell into the Caspian Sea, the dried up bed of which is the present Uzboy. The river Arzas, it is said in the Book, flowed out of the Blue Sea, i. e. out of the Aral, which is evidently false, but the error took place clearly from confounding the lake basin of Sary-Kamysh with that of the Aral, which often occurs in oriental writers, who were in a position to be and were in reality incomparably better acquainted with Central Asia than the Russians.

In another passage of the Book it is said: «And opposite «to the town of Bokhara, at a distance of 170 versts, a ri-» «ver flowed out of the lake Ugus, in our tongue Bull, into» «the Sea of Khvalim; its course is a thousand versts».

Under the name of Ugus is to be understood the Oxus of the ancients or the present Amu-Darya, which passes at a distance ot about 120 versts to the west of Bokhara. The distance from the Amu-Darya opposite Bokhara to the mouth of the Uzboy in the Bala-Ishem depression, which was filled up with water at that time, approaches somewhat to the distance given in the Book.

Thus in the Book of the Great Map, the Amu-Darya is mentioned under the names of Arzas and Ugus, just as the ancient writers knew this river under the similar names of Araxes and Oxus. But, as it is nowhere said in the Book that these names mean the same river, it must be concluded that this information; true in itself, was taken from different sources, which did not quite agree with each other. At any rate, the Book of the Great Map mentions only one mouth of the Amu-Darya, namely that in the Caspian Sea, which is the most essential fact for us.

Of the sources of the XVI Cent., relating to the course of the Amu-Darya, special attention should be paid to the Great Encyclopedia Jyhan-Numa, compiled in 1650 by Hadji-Khalifa. Although, according to more authentic sources which will be spoken of below, the stream of water by the Uzboy must at that time have ceased to exist, yet the information contained in this work is important, as an echo from the preceding epoch. This is what is said in it about the Amu-Darya: «One of the arms» «of the Jyhun, having passed the capital of Khovarezm, forms a wa-» «terfall, where it is precipitated with such a terrible sound, that it is» «heard at a distance of two parasangs. According to the testimony» «of Hamdallah this arm falls into the Caspian Sea».

This evidently refers to the present Uzboy, in which immense cataracts were found and which has its mouth in the Gulf of Balkhan in the Caspian Sea. The author of the Jyhan-Numa, describing the course of the Amu-Darya, says among other things, that this river below the pass of the «Lion's Jaw» (Tiuya Muyun) is lost under ground and afterwards reappears on the surface. This evidence, in connexion with other similar testimonies, serves as a confirmation of the fact, that the Amu-Darya had cut a subterranean passage through the Sheikh-Djeili range.

Finally, there is yet one more confirmation of the fact, that at that time the waters of the Amu-Darya had on the one hand a mouth in the Aral Sea and on the other in the Caspian, namely in Chinese sources. According to Chinese official information, the river Amu-Darya divided into arms, of which one ran north and the other southwest, their mouths being in two diffe rent basins.

But the stream of water by the present Uzboy, which was formed in the XIV Cent., began with time somewhat to decline, and evidently ceased to exist about the middle of the XVI Cent. We meet with trustworthy information on this matter in the English traveller Jenkinson. In the middle of the XVI Cent., the English company of Merchant Adventurers despatched into Central Asia Anthony Jenkinson, for the purpose of becoming acquainted with the commerce of those countries and of seeking a market for the sale of English goods. After landing on the eastern shore of the Caspian Sea, in the Gulf of the Czesarevich, Jenkinson started for Khiva on camels with a caravan and, after a journey of twenty days through a waterless desert, reached a gulf, in his opinion, part of the Caspian Sea. In this gulf he found such excellent drinking water that he halted there a whole day... «Note», says he, «that in times past, there did fall into this gulf the great» «river Oxus, which... and now cometh not so far but falleth into» «another river called the Ardok, which runneth toward the north» «and consumeth himself in the ground, passing under ground» «above five hundred mile and then issueth out again and falleth into» «the lake of Kitay»..

Thus Jenkinson, after a journey of twenty days came to the gulf of some sweet water lake. This lake was so large, that he took it for a gulf of the Caspian Sea. There can be no doubt but that this was the lake basin of Sary-Kamysh. It is remarkable, that he states categorically, that in ancient times the Oxus fell into this lake which, when he was there, i. e. in autumn, no longer reached the lake. The Oxus, as Jenkinson states, fell in those times into another river, the Ardok.

It is evident, that by the Ardok must be understood the chief channel of the Amu-Darya, within the limits of Khovarezm, which has a mouth in the Aral Sea. The fact that the Ardok, below the Oxus, i. e. the Amu-Darya, is lost under ground and afterwards reappears again on the surface, only confirms the testimony of Hamdallah and Hadji-Khalifa on the formation by the river of a subterranean passage through the Sheikh-Djeili range. The lake of Kitay into which, according to the observation of Jenkinson, the river Syr-Darya falls, is of course the Aral Sea.

Farther Jenkinson says: «We, having refreshed ourselves at the» «foresaide gulf, departed thence the 4-th. day of October and the» «7-th day arrived at a castle called Sellizure»... «The castle of Selli-» «zure is situated upon a high hill, where the king called the Can» «lyeth. The southern part of this castle is lowe land but very fruit-» «full, where growe many good fruites... the water, that serveths all» «that country, is drawn by ditches out of the river Oxus unto the» «great distruction of the said river, for which cause it falleth not» «into the Caspian Sea, as it hath done in times past, and in shor» «time all this land is like to be distroied and to become a wilder-» «ness for want of water, when the river of Oxus shall faile».

The castle of Sellizure, situated, as Jenkinson says, on a height, southwards of which extends a low land, which is irrigated by water drawn out of the Amu-Darya, is apparently nothing else but the ancient town Kariat-Farategin situated by the present ruins of Deu-Kesken.

Consequently, during the sojourn of Jenkinson in these places, that is in the middle of the XVI Cent., the ordinary low waters of the Kunya-Darya went only as for as Deu-Kesken and no longer reached the lake of Sary-Kamysh and it was possible to foresee the time, when they would not reach even this lake and when this flourishing country would be reduced to a waterless waste. To be sure, to speak more precisely, it should be said, that in the time of Jenkinson the water did not reach the Sary-Kamysh only in late autumn, at low water, and there can be no doubt but that at the flood a very considerable quantity of water still passed in this di rection, but in any case it is doubtful, whether this water could be sufficient to sustain a continuous current below the Sary-Kamysh, by the Uzboy.

Thus, in this conscientious English traveller who, thanks to a fortunate concurrence of circumstances, visited Central Asia just at the point between the epochs of the existence of the communication between the Amu-Darya and the Caspian Sea and of the discontinuance of this communication, we find a very important witness in favour of the fact, that the flow of water by the Kunya-Darya, which was the chief affluent of the lake basin of Sary-Kamysh, existed a little before his arrival, but then already was beginning to disappear. The account of his journey breathes such a truthfulness, that there can scarcely be a doubt of the correctness of the facts, which he communicates as observed by himself.

Another in the highest degree remarkable proof of the changes, which occured in the course of the Amu-Darya, is to be found in the «History of the Mongols and the Tartars», a book written by a man, who during twenty years was autocrat of Khovarezm. Abul-Ghazi-Behadur-Khan was born in Urgench 47 years after the Englishman Jenkinson visited this town and gave the first account of the discontinuance of the communication of the Amu-Darya with the Caspian Sea. About the former course of the Amu-Darya Abul-Ghazi relates the following: «in those times (that is» «at the accession to the throne of Sofian-Khan, about 1525) all» «the road from Kunya-Urgench to the Abul-Khan (the» «Balkhan Mountains) on the south east coast of the Cas-» «pian was covered with «auls»; for the Amu, after pass-» «ing under the walls of Urgench flowed down to an» «outlet at Ogurcha (in the Balkhan Bay) and both banks» «of the river to its mouth were cultivated and covered» «with vineyards and orchards and were very populous» «and in a most flourishing condition. In the spring the» «inhabitants retired to the mountains and drove off their» «herds during the time of gnats and flies to the wells, si-»

«tuated at a distance of one or two days journey from»
«the river, to which they returned only when the insects»
«disappeared. This whole country was very populous and»
«in a most flourishing condition».

We find thus in this quotation the most positive statement of the fact, that the communication of the Amu-Darya with the Caspian Sea still existed in the time of Sofian-Khan, that is in the beginning of the XVI century, and that at that time the country on both banks of this river was densely inhabited and in a flourishing state. In another passage of his book Abul-Ghazi speaks as definitely and precisely of the change of the course of the Amu-Darya «I was born in the state of Urgench in 1014 of the» «Hegira (July 1605). Thirty years before my birth», (that is about 1575) «the Amu-Darya, cutting itself a channel from the» «place called Kara-Oigur-Tokai, above Khast-Minaressi to Tuk-» «Kalassi, flowed into the Sea of Syr (Sea of Aral), in conse-» «quence of which the neighbourhood became changed into a desert».

According to this passage the time, when the flow of water by the Kunya-Darya ceased, agrees with the preceding quotation and nearly approaches the time of Jenkinson's expedition. Khast-Minaressi, where the new bed of the Amu-Darya began, must be assumed to be somewhere near Khiva and Hazarasp, near the canal Yarmysh, as follows from the combination of facts given by Abul-Ghazi. Tuk-Kalassi, or simply Tuk, was a castle on the right bank of the river to the north of Kyat and at a short distance from Urgench probably in the vicinity of the castle Nukus. From these quotations, it is evident that the Amu-Darya flowed up to the year 1575 in a south-western direction close to Urgench then reached the hills of Balkhan and fell into the Caspian Sea; but after this year it went to the castle of Tuk and fell into the Aral. In this matter the testimony of Abul-Ghazi, a man, who lived a little after the above mentioned change of course, who was well acquainted with the topography and history of the country, an impartial man and not a mere bookworm, deserves the most complete confidence. This testimony is in the highest degree important, as a new confirmation of the changes, which occurred in the lower course of the Amu-Darya.

There are in the book of Abul-Ghazi yet indirect indications, that at the beginning of the XVII century the level of the Sea of Aral was lower than at the present time, as may be judged from the fact that in the limits of the delta of the Amu-Darya, there were immense cultivated areas.

«This whole locality» (in the present delta of the Amu-Darya), «sowed only with wheat», says Abul-Ghazi, «was so» «vast, that a mounted man could not ride round it in» «ten days.»

It is evident, that if the level of the Sea of Aral stood at the same height as it does at present, such a great space of cultivated land could not have existed, where is the present delta of the Amu-Darya, because in consequence of the pressure of the sea there would have been great inundations and marshes, just as it is at present in the same place. Beside this, in the book of Abul-Ghazi there are indications, that the fall of the canals carried through the present delta of the Amu-Darya was so considerable, that their banks were violently washed away by the strength of the current, and a small canal, dug through the delta, became in the course of a few years wider than the flight of an arrow. This fact is perfectly intelligible. The lower level in the Aral must have increased the fall of water in the arms, which had their mouths in it, in consequence of which an increased washing away of the banks in the arms must take place as well as in the artificial canals. Both of these facts, the extent of cultivated land in the delta and the washing away of the banks of the canals, show indirectly that formerly the quantity of water, which fell into the Sea of Aral, was smaller than subsequently, which could have occurred only in consequence of the interruption of the former communication between the Amu-Darya and the Caspian Sea. Thus the testimonies of such impartial and conscientious writers, as Jenkinson and Abul-Ghazi, prove in the most positive manner, that the communication between the Amu-Darya, and the Caspian Sea, which was reestablished in the XIII century, ceased again in the latter half of the XVI century, and that the river from that time took the direction, which it has at the present time.

What were the causes, which called forth this phenomenon, is a question, which we must examine more in detail.

According to the information of Abul-Ghazi, a stream of water by the Uzboy existed in the reign of Sofian Khan in 1525. Consequently a considerable quantity of water must have fallen into the lake Sary-Kamysh. In 1558, at the time of Jenkinson's travels, there was already not enough water for the irrigation of the fields-along the Kunya-Darya. Finally in 1575, according to the same Abul-Ghazi, a revolution took place, in consequence of which the neighbourhood of Urgench was reduced to a desert. It is to be noted, that this author names this very year, and not a certain interval of time, which, properly speaking, he should have done, if the ceasing of the current had taken place gradually. Besides this he says, that at that time, «the river Amu cut a channel for itself». This means, that the chief channel, by which the river began to flow from that moment, was not formely its chief bed. Thus, the chief stream of the river passed from one arm into another. The river Amu-Darya even now is divided into arms, which often change their course, and in consequence of the easy washing away of the banks and of the abundance of drifts these changes take place very rapidly. Two or three floods are sufficient to force the river to take the direction of the former bank; and in the place of the recently existing bed for a new bank to be formed. There is no ground to suppose, that this property of the Amu-Darya manifested itself in the XVI century less intensely, than at the present time, and therefore the increasing of the flow in certain arms of the Amu-Darya at the expense of the others contains nothing impossible. In the given case there is grouud to suppose, that the increase of the current took place in the eastern arms of the Amu-Darya at the expense of the western arms and not vice versa. It was mentioned above, that a part of the waters of the Amu-Darya, having reached by its arms or irrigating canals the range Sheikh-Djeili, passed through them by subterranean passages, as is mentioned by different authors, beginning with Edrissi.

It must be admitted, that at that time, that is in the latter half of the XVI century, occurred the complete breaking through of this ridge; the limestone rocks, hanging over the water, broke down and the river here formed itself an open bed.

Some facts can be brought forward in favour of this hypothesis. Firstly, the latest travallers, as far as we know, make no mention of the Amu-Darya or any of its branches being lost under ground, while the authors of the preceding epoch, as we have seen, often mentioned it.

Secondly, a part of the present chief bed of the river, from the town of Gurlen to the beginning of the Lauzan canal bears the character of recent formation, indicated not only by the lime rocks of Sheikh-Djeili, which descend perpendicularly to the very river, but also by the character of the left bank of the Amu-Darya, which beyond these hills is generally lower than in other places. This must be attributed to the circumstance that on this bank alluvial drifts had no time to be deposited in a sufficient quantity.

In favour of the recent existence of the current in this part, speaks the considerable decrease of the fall and the slowing of the current in the lower parts of the bed below the town of Kipchak. It is evident, that under such circumstances an energetic formation of deposits must have taken place, at this point, which with time must have raised the bed and increased the fall. If this did not take place before the present time, it was only because the river began to flow by the present bed comparativly not long ago. Finally in this part are met with lakes, as for instance Kosh-Tube, Meili-Kul and others, which have not yet completely separated from the Amu-Darya and form remains of former local hollows. And as the arms of the Amu-Darya, which fed the lake of Sary-Kamysh, parted from the old arm, which apparently passed a little more to the west of the hill Yumur-Tau, the current by those arms began evidently to decrease rapidly with the perforation of the Sheikh-Djeili range, and the removal of the chief bed thither. How rapid was the increase of the right new bed at the expense of the old bed, may be judged from the fact that at the present time there are no traces left of it. At the same time the new banks of the Amu-Darya below the Sheikh-Djeili range have not even yet succeeded in forming themelves completely.

Thus, notwithstanding the absence of positive facts, it may be assumed, that the definitive perforation of the Sheikh-Djeili range took place in the latter half of the XVI century, and that this circumstance was the first cause of the weakening of the current in the Kunya-Darya, particularly in its upper part, and of the obstruction of the arms, that had separated from the left bank of the river.

But by the perforation of the Sheikh-Djeili range it is impossible to completely explain such a complicated phenomenon, as the cessation of the flowing of the waters of the Amu-Darya, not only into the Caspian Sea, but even into the lake of Sary-Kamysh. Above, in the description of the canals in Khovarezm under the dominion of the Arabs, there was mention of the Buh canal which separated from the Jyhun (Amu-Darya) already in the neighbourhood of Djordjania. This canal joined the Wadak (the Kunya-Darya) at a distance of one day's journey from that town, so that it must have passed in the direction of the present Lauzan. This direction forms, so to say, a prolongation of the chief bed of the Amu-Darya and therefore its course must have tended to take the direction of this canal. Besides this from the levels taken it appeared, that the average fall of the Lauzan corresponds with that of the chief bed, lying above, of the Amu-Darya and surpasses that of the lower part of the bed, which is expressed by the increased washing away of this canal every time the waters of the Amu-Darya are let into it. There is no reason to suppose, that the fall of the Buh was less considerable than that of the present Lauzan. This circumstance must have contributed still more to the tendency of the waters of the Amu-Darya to this canal. Besides this, the decrease of the current in the upper part of the Wadak (the Kunya-Darya), which took place in consequence of the perforation of the Sheikh-Djeili range, lowering the level of the water in its lower parts and thus increasing the fall in the Buh, must have still further increased the current in this direction.

The question suggests itself why, notwithstanding all favourable conditions, the river Amu-Darya, after having cut its way through the Sheikh-Djeili, did not direct itself by the Buh (the Lauzan) and cut itself in this direction a new bed to the Sary-Kamysh. What were the causes, which forced it to choose the less convenient course to the Aral? During the explorations of the Imperial Russian Expedition, the inhabitants explained, that the canal Lauzan was so named after a citizen, who first for the irrigation of his fields dug a little ditch from the Amu-Darya which, in consequence of special favourable conditions, was afterwards enlarged by the current itself to its present dimensions. But farther washing away did not take place, because the Khivan Khans, from political views, did not wish to let too much water flow to the Turkomans; so they stopped up the bed of the Lauzan at its exit out of the Amu-Darya with solid dams. A fort called Bend was built for the purpose of watching the condition of the dams erected, the ruins of which still exist.

In the Lauzan canal, besides the Bend dam, at its exit out of the Amu-Darya, were found three other dams, which stopped up its current. The construction of the dam Tash-Bend at the outlet of the Lauzan into the Kunya-Darya as well as that of the Bend dam, at its exit out of the Amu-Darya were remarkably solid. According to the explanations of the natives, the construction of these dams was also called forth by political considerations, with the intention of depriving of water the inhabitants, living lower down the river, who were on hostile terms with the Khivan Khan. These explanations of the reasons for the erection of dams for arresting the waters of the river Amu-Darya from flowing westward, find confirmation in a whole series of ancient traditions. Already the scouts of Prince Bekovich Cherkassky, as will appear below, sent by him to Khiva for the purpose of getting information relative to the old bed of the Amu-Darya, brought the intelligence, that the Amu-Darya might be directed into the Caspian Sea, for which purpose it was only necessary to destroy the dam, erected for security from the Russians in its old bed. This dam, which these explorers had seen themselves, was situated, as appears from their report, at the exit of the former Buh or the present Lauzan. Moreover, nearly all the traditions from the beginning of the XVIII Cent., quoted by Florio Beneveni, Blankennagel, Rychkov, Rukavkin, Velichko, Eichwald and others, also attribute to the same cause the turn, which took place in the course of the Amu-Darya, and explain the construction of dams by local Khans as the result of political views, namely for the purpose of subjecting to themselves their restless nomad neighbours, living lower down the course of the former river bed, or to secure themselves from the attacks of Russian volunteers, or simply out of fear of attracting the warlike steppe tribes. It cannot possibly be admitted that all these traditions, so conformable among each other, in defining the causes of the interruption of the communication of the Amu-Darya with the Caspian Sea, are deprived of all foundation, the more as they agree with the most modern explorations of the topographical conditions of the Lauzan, which show that, if there had not been any artificial impediments on it, such as dams, the waters of the Amu-Darya not only would have directed their course by it, but even would have washed themselves a bed of suitable dimensions in that direction. It must still be noted, that the definite turn of the course of the Amu-Darya into the Aral took place just at the time of the domination of the Uzbeks in Central Asia. Despising peaceable occupations, they constantly devastated and ruined by their intestine wars and rapacity the once splendid and flourishing Khovarezm. This was the cause of the political weakness of the Uzbeks. Two years before the birth of Abul-Ghazi the Cossacks of Yaik (Ural) took and plundered Urgench, which proves the complete political helplesness of Khovarezm at that period. The Turkomans, before whom Persia trembled, were of course more dangerous for the Uzbeks than the Cossacks of Yaik. If the Turkomans recognised the supremacy of Khovarezm, as long as it was rich and powerful, when they became stronger, the only means to keep them in submission remaining to the Uzbeks, was the depriving them of water. And as to the circumstance, that the results of such a measure might have been the reduction of the flourishing neighbourhood of Urgench to a desert, that apparently troubled the hardy nomads very little.

Thus from all the above we must absolutely arrive at the conclusion, that the second and most important cause of the interruption of the course of the waters of the Amu-Darya into the Caspian Sea was the erection, from political views, of dams in the former Buh, or present Lauzan.

But in this complex phenomenon, apparently played a part a third factor, namely the erection of dams from economical views for facilitating irrigation, with which object, as we have seen, was erected in the X century the dam' across the present Kunya-Darya, near the town of Djordjania, causing the interruption of the current by the Uzboy. This cause Jenkinson also indicates in his notes of his travels: «As this river (the Oxus-Kunya-Darya) dried up in many places in «consequence of the summer heats, the Turkomans imagined, that by» «means of erecting a dam at its outlet (in the Sary-Kamysh) they» «could secure the higher part of the river from the disadvantage of» «want of water, but the result was just the contrary, because the» «current of water, already not being strong enough to remove the» «sand, which was continually carried into the stream by the desert» «winds, the river bed was finally filled up with sand, so that there» «remain at present only some few traces of it to be seen yet in» «the vicinity of the sea (Sary-Kamysh)».

At the present time, on a more detailed exploration of the old beds of the Amu-Darya, joining it with the basin of the lake Sary-Kamysh, it has been discovered, that these beds are in many places barred with dams. In the Kunya-Darya a whole series of such dams was discovered, Kizil-Takyr, near the town of that name, the dam of Shamrat, at the exit of the canal bearing the same name, then the dams of Medemi, Tadjik-Bend, Ushak-Bend, Salak-Bend and Egen-Klych, situated near the ruins of Deu-Kesken. In the higher part of the Daudan, in the Khivan Oasis, were discovered the dams Tiachi-Bend and Pitnek-Bend, and in the lower part of this bed, in the desert, were found yet four dams, of which the lowest bears the name of Kum-Bugut. These dams, according to the words of the natives, were constructed for the purpose of raising the water in the bed to irrigate the neighbouring fields, without which it would have been impossible, because the quantity of water which flowed through the old beds was not sufficient to fill up the whole bed, and the level of the water must have been considerably lower than the surrounding banks. On the basis of these facts, the conclusion must be drawn, that as he quantity of water, flowing by the old beds of the river Amu-Darya, the Kunya-Darya and the Daudan, was considerably diminished in consequence of the perforation of the Sheikh-Djeili range and of the erection from political views, of dams in the present Lauzan and as the water already could no longer fill the whole of these beds, a series of dams was erected, with economical views, in these beds for the purpose of satisfying the needs of irrigation of the adjacent localities, through which these dams arrested the water in the beds and favoured its enormous consumption in evaporation and filtration.

Thus the interruption of the communication of the river Amu-Darya with the Caspian Sea, which occurred in the latter half of the XVI Cent. may be explained by three causes:

1) By the perforation of the Sheikh-Djeili range by the Amu-Darya, which diminished the quantity of water flowing in the Kunya-Darya.

2) By the erection of dams, with political views, in the system of the former Buh, or the present Lauzan, which hindered the formation of à new bed of the Amu-Darya to the Sary-Kamysh.

3) By the erection of dams, from economical views, namely for facilitating the irrigation of the neighbouring localities, in the present old bed of the Kunya-Darya and the Daudan.

To the united action of all these factors must be attributed the circumstance that the current of water by the old beds not only decreased, but even ceased altogether, and that the basin of the lake Sary-Kamysh, being deprived of the stream of waters from the Amu-Darya, began to gradually diminish in volume and to dry up. At the present time, only insignificant traces remain of it in the form of the Sary-Kamysh lakes of to-day. On the other hand in the X century, when the sole factor of the interruption of the communication of the Amu-Darya with the Caspian Sea was the erection of the dam in the old bed of the river, the basin of Lake Sary-Kamysh receiving a considerable quantity of water from the Amu-Darya, continued to exist, and had, as we have already seen, even its special name.

In any case, it must be supposed that the change of the course of the river Amu-Darya to the Sea of Aral did not take place completely in the XVI century, and that the river, from time to time, washed away the dams constructed and the drifts accumulated and flowed as before by the western arms into the basin of the Sary-Kamysh. As to the XIX century, there is positive information that this took place about the year 1812, in 1834, 1849, 1850, 1857, and finally in 1878, when the water broke through and reached the lake of Sary-Kamysh. The first information relative to the old beds of the Amu-Darya was obtained during the Khivan expedition of Prince Bekovich Cherkassky. Commercial relations, between the merchants of Astrakhan and the nomad Turkomans inhabiting the steppes of Central Asia, existed long ago, and Cape Tiub-Karagan at Mangyshlak was visited by them with that object. With one of such commercial companies, a noble Turkoman, Hodja Nefes, came to Astrakhan in 1713 and announced to Prince Samanov, who resided in that town, that he had to make known a matter of great political importance to the Emperor. Having accompanied Nefes to St. Petersburgh Prince Samanov was, with him, presented to the Emperor. The revelation of the Turkoman Nefes consisted in the statement that, in the country lying on the river Amu sand gold is obtained, and that although that river, which used to fall into the Caspian Sea, was for sake of security from the Russians led off by the Uzbeks or Khivans into the Sea of Aral, it might be turned back into its old bed by cutting through the dam, in which undertaking the Turkomans would assist the Russians. The intelligence of the obtaining of gold sand in the upper part of the Amu-Darya was confirmed by the Voyevod Prince Gagarin, and the Khivan ambassador Ashur-Bek. Peter I, whose hands were free after the termination of the war with Charles XII and the Turks, resolved to take advantage of the occasion for the consolidation of his power in Central Asia. But before taking any more resolute action to open new sources of wealth for his country in the gold fields of Central Asia, and especially in the reestablishment of the relations that had once existed with this country, it was necessary to acertain how well founded was the information communicated by Nefes. For this purpose, Peter the Great first equipped an exploring expedition to Khiva under the command of Prince Bekovich Cherkassky.

By an Ukase in the name of His Majesty the Czar the above named prince was directed to go to Khiva with greetings to the Khanate, and thence to go to Bokhara to the Khan under pretext of some commercial transaction, and in reality to make inquiries about the town of Yirken (Yarkend, where the auriferous sands should be), how far it is situated from the Caspian Sea, and if there are not any rivers from there, or if not from that place itself, at least in the vicinity of the Caspian Sea.

This expedition, having started in 1715 on board ship from Astrakhan and following the sea coast, reached Cape Tiub-Karagan in Mangyshlak. Having landed with the whole force, and having learned from the local Turkomans that it was only necessary to cut a canal of about 20 versts for the river Amu-Darya to flow again into the Caspian Sea, Prince Cherkassky ordered two Astrakhan nobles, N. Fedorov and Iv. Zvansky, to examine the locality under the direction of the Turkoman Hodja Nefes, and to return to Krasny Vody (Krasnovodsk), whither the remaining part of the expedition started by sea. The scouts proceeded by the caravan road to Khiva and, after a journey of 17 days, reached the river Karagach, across which a dam was erected, preventing the waters of the Amu-Darya from flowing to the west. In Fedorov's report, it is said that, two versts or so before reaching the river Darya, Nefes led them to an earthen rampart, which was about 3 feet high, 21 feet wide and 3½ miles long; the river Darya was at that time at high-water and reached this bank. And the Turkoman Nefes said that, if this bank and the steppe behind it as far as the valley that is there be cut through, the water will flow from the river Darya into the Caspian Sea. And Hodja Nefes says in his account as follows. «The river Karagach, which issued in the Khivan territory from the» «Darya, flowed from ancient times through the steppe and between» «the hills and fell at its mouth into the Caspian Sea, three days' jour-» «ney above Krasny Vody (Krasnovodsk), and this river was dammed» «up by the Khivans in ancient times, who erected a dam at about» «two versts from the river Darya. In consequence of this there is no» «longer any stream from the Karagach intothe sea beyond this dam,» «and the river Karagach returned back into the river Darya».

The river Karagach, across which the dam was built interrupting its stream, was of course nothing else but the present Lauzan, which forms a connecting link between the Amu-Darya and its old bed the Kunya-Darya. The dam, arresting the water in the chief bed of the Amu-Darya and situated at the exit of the river Karagach from the Amu-Darya, must have been somewhere in the neighbourhood of the present dam Bent at the outfall of the Lauzan. Prince Cherkassky's scouts, having started from the river Karagach along the old bed of the Amu-Darya having its mouth in the Caspian Sea, quitted this bed at the site of Ata-Ibrahim, and reached by a direct road Krasny Vody, where Prince Cherkassky already was.

Having returned to Astrakhan, Prince Cherkassky reported to the Emperor the results of his reconnaissances. Having received this report, Peter I summoned the Prince to come to him and, at the personal explanation, Prince Cherkassky succeeded in raising great hopes for the success of a new expedition. On the 14. of February 1716, the following Ukase of His Majesty was given to the Senate:

«Gentlemen of the Senate! As We have sent Captain Prince» «Cherkassky back thither, whence he came, and given him instructions» «on what he shall do there, do you carry out without delay, what he» «shall require of you, according to all the points of these instructions» «and even beyond them».

With this Ukase, Prince Cherkassky was given a special order, consisting of thirteen points in which, among other things, it is said:

«Commanding the port, where the mouth of the Amu-Darya» «once was, a fort is to be built for about a thousand men, as re-» «quested by the Khivan ambassador».

«To go to the Khan of Khiva as Ambassador, and to take the» «route along that river, and to examine carefully the course of that» «river as well as the dams; if it is possible, to turn this water back into» «the old bed also at the same time to shut up the other mouths,» «which are in the Sea of Aral; how many men are needed for this work».

«To look out for a spot near the dam, or where convenient, on» «the present Amu-Darya, for building a fort secretly, and if it will be» «possible, to construct there another town».

«To ask the Khan of Khiva to send his men, with two of our» «own, by water on the river Syr-Darya to the town Irket to exa-» «mine the gold».

«Also to ask of him boats, and to send by them a merchant» «by the Amu-Darya to India with orders to sail up the river as» «far as the boats can go, and thence to travel into India, noting all» «the rivers and lakes and describing the water and land route and» «especially the water route to India by this or other rivers, and to» «return from India by the same route or, if he should hear in India» «of a still better route to the Caspian Sea, to return by it, and describe it».

In 1716, Prince Bekovich Cherkassky sailed from Astrakhan with a detachment of more them 6000 men in order to build forts on the coast of the Caspian Sea; and in the following year was undertaken the well known expedition into Khiva from fort Guriev by land through the steppes. The deplorable result of this expedition destroyed at one stroke the comprehensive plans of the Great Emperor for turning the river Amu-Darya into the Caspian Sea, and the consolidation of Russian dominion in Central Asia, and at the same time arrested for many years further attempts in this direction. Only at the end of our century, when the Russians established themselves finally in Central Asia, did it appear possible to continue the work of the Great Reformer, and to proceed again to a systematic exploration of these river beds. For the whole interval of time from Peter the Great to the seventies, information upon these beds is casual and fragmentary. Individual travellers penetrated from time to time into the depths of Central Asia, but they were able to see and observe these beds only at certain spots, where they were crossed by caravan roads. The information obtained upon these beds must, therefore be exceedingly incomplete and, fragmentary and the sole proofs of the existence in former times of a current in this direction are traditions and relations of the native population who, be it remarked, affirm this with one voice.

Thus, in the middle of the XVIII century, the English, merchants *Thompson* and *Hogg* and a Samara merchant *Danilo Rukavkin*, on their way through the Kirgiz steppes to Khiva, crossed the old bed of the Amu-Darya, the present Kunya-Darya. On the other hand, *Woodcrof*, visited the coasts of the Caspian Sea at the Gulf of Balkhan. At the end of the same century, Dr. *Blankennagel*, on his way to Khiva and back, had the opportunity of seeing the old bed of the Amu-Darya, the present Kunya-Darya. In 1819, *Ponomarev* and *Muraviev* made the journey from Krasnovodsk to Khiva and crossed the Kunya-Darya where it falls into the lake of Sary-Kamysh. On their way back, they crossed the old bed of the Uzboy. In 1825, *Berg* determined exactly the distance between the Caspian and the Sea of Aral, and the difference of their levels with the aid of barometrical observations. In the same year, *Eichwald* made his scientific expedition to the Gulf of Balkhan, and in the year 1836 this gulf was visited by *Karelin*. In 1839 and 1840, the Englishmen Captains *James Abbot* and *Shakespeare*, on their return from Khiva to the fort Novo-Alexandrovsk, crossed the old bed of the Kunya-Darya, In 1842, an embassy was sent to Khiva under the command of Lieutenant Colonel *Danilevsky*, who was accompanied by the naturalis, Basiner. This embassy crossed the Kunya-Darya at Kunya-Urgench, and on its way back sailed across the Gulf of Aibugir Both members of this embassy left excellent descriptions of their voyage. In 1851, the Persian ambassador *Riza-Kuly-Khan*, on his way to Khiva, crossed the old bed of the Uzboy. Finally in 1863, *Vambery*, under the disguise of a mussulman pilgrim or hodja, made his remarkable journey and during it twice crossed the Uzboy. But, up to the present, all explorations of the Amu-Darya's old beds bore a casual character. An insuperable obstacle to a more detailed examination of these beds was presented not only by the severe natural conditions of the climate of the Turkoman steppes, but also by the warlike and predatory character of the population. It became possible to begin a systematic exploration, only after the consolidation of the

Russian influence in Central Asia. When, in 1869, Russian troops occupied Krasnovodsk and founded a fortress there, then Russia became finally established in the Transcaspian steppes, and there was immediately undertaken a whole series of expeditions for the systematic exploration of the old bed of the Amu-Darya, the Uzboy.

The Viceroy of the Caucasus, HIS IMPERIAL HIGHNESS THE GRAND DUKE MIKHAIL NIKOLAEVICH, took the matter under his immediate protection, constantly supported it with his attention, and powerfully contributed to the successful course of the investigations of the ancient bed of the Amu-Darya.

At the end of the year 1870, an expedition was equipped under the command of colonel Stolietov and sent into the depths of the Turkoman steppes to Fort Kisyl-Arvat. Colonel *Stebnitsky*, who accompanied the expedition, explored the lower part of this river bed as far as the Aidin wells. At the same time, a survey of the locality, the determination of latitudes and longitudes and barometrical altitudes of different points were taken.

Lieutenant Colonel *Markozov* undertook in the autumn and winter of the year 1871 three reconnaissances in the steppes, of which two had a direct reference to the old beds of the Amu-Darya. One of them was directed straight through the steppes to Lake Sary-Kamysh, and reached the wells of Dekcha on the Kunya-Darya at a distance of 33 versts from its mouth. The other, which was undertaken to the Uzboy, explored this river bed beyond the wells of Aidin as far as Lake Topiatan. The investigations, during the reconnaissances, consisted of the making of route surveys of the localities.

In 1872, colonel Stebnitsky, attached to the force commanded by colonel Markozov, continued the investigation of the Uzboy from the Lake of Topiatan to the wells of Igdy. During this expedition, the latitude and longitude of nine points on the Uzboy were determined, and a detailed survey was made of the whole of the Uzboy already investigated.

But, until the year 1873, all these explorations of the old beds of the river Amu-Darya were made only from the side of the Gulf of Balkhan and in no way concerned the limits of the Khivan Khanate. It was only possible to fill up this blank after the Khivan expedition, which extended the Russian dominions in Central Asia and opened access to the little known lands of the Amu-Darya.

In 1873, after the taking of Khiva, Adjutant General Kauffman commanding the active troops, at the request of Colonel Glukhovskoy authorised him to take advantage of the march of the Orenburg detachment through the western part of the Khanate to explore the water system of the Khanate in this direction. For this purpose, an expedition was organized under the command of Colonel Glukhovskoy, since known under the name of the *Urun-Darya Expedition.* It marched out, together with the Orenburg detachment, from Khiva the 19. of July, and immediately began to make surveys and gather information about the old beds of the Amu-Darya. After having crossed the canal Khazavat, the expedition came across one of these old beds, called Daudan. Great was the effect, when the members of the expedition perceived running water in this bed, which is from 1750 to 2800 feet wide. The detachment partly forded the bed, partly passed by a bridge thrown over one of its arms. On the whole space, from the village Igdyr-Kala to the town of Tashauz, this bed had approximatively the above stated width. In the town of Tashauz was formed a surveying party, which started in two boats up the canal Shakh-Abat and descended by a connecting canal into the old bed of the Amu-Darya, by which it continued until the stream of water ended.

In consequence of the importance of the exploration of the Kunya-Darya, General Kauffman authorised the expedition to proceed, not as was originally intended to Khodjeili, but to go straight to Kunya-Urgench. The whole way, the surveying party followed the bed, while the detachment marched straight along the road. When the canal Shamrat was reached, a farther march along the river bed proved to be perfectly impossible, because the party was compelled to force its way through an impassable thicket of saksaul, tamarisks and thorns, out of which it succeeded in getting only with difficulty. After having joined the detachment, the party continued its way with it to Kunya-Urgench. In this manner the western part of the Khivan Khanate from Khiva to Kunya-Urgench was for the first time traversed by Russian troops, and for the first time authentic information was gained on the situation and condition of the ancient beds of the Amu-Darya, Daudan and Kunya-Darya. These beds proved to be sufficiently abundant in water and to be fed only by the neighbouring canals, but not to have any direct communication with the Amu-Darya. The Daudan proved to be particularly wide and full of water. After reaching Tashauz, it divides into two arms. The left passes by Izmukshir and the right by Ilialy; both, after flowing a certain distance, are lost in the steppes. The Kunya-Darya, beyond Kunya-Urgench, bore the name of Urun-Darya on old maps. The natives affirmed that, not more than eleven years ago, the water by the Urun-Darya reached the Sary-Kamysh. On the Amu-Darya and its old beds, a whole system of dams was found, barring the water, and forcing it to flow out of the chief beds into various canals in certain directions.

After the arrival of the detachment at Kunya-Urgench, the Expedition of Urun-Daria marched farther to reconnoitre the old bed of the Amu-Darya, from Kunya-Urgench to the lakes of Sary-Kamysh. The exigencies of a time of war compelled the expedition to carry out its investigations rapidly.

Early in the morning of the 1. of July, the expedition started from Kunya-Urgench for the wells of Nazar-Bi, near the dam of Salak-Bent. The road ran along the right bank of the Urun-Darya or the Kunya-Darya. The locality was sandy, the descents and ascents steep, tamarisk and other shrubs covered the neighbouring country on both sides of the road. The baggage train moved with difficulty, especially through the sand, into which the wheels sank deep. The bushes and tamarisk trees occasionally arrested the movemen of the train by catching the harness and making the driving of three horses abreast impossible, so that it became necessary to unharness the third horse. Some versts from the bivouac, many ruin of kishlaks or villages, and a quantity of dried up aryks or ditches witnessed that, not long since, there existed here both population and agriculture. The bivouac was selected two versts from the Salak-Bent dam. The march made of about 45 versts was very heavy. In consequence of the long march, and the complete uncertainty as to the possibility of moving along the old river bed, its exploration from Kunya-Urgench to Salak-Bent was deferred to the return journey. To carry out a survey, taking levels, proved equally impossible, both on account of the necessity for the rapid movement of the detachment and in consequence of the broken ground, covered with a dense and high brushwood of tamarisk.

The 2. of July, early in the morning, the Expedition marched on farther in two columns. Near to the Salak-Bent dam, on both banks of the river bed, were found dry flanking canals, which formerly served for the passage of the water. Farther on in the direction of Egen-Klych, ruins of kishlaks on both banks of the bed were met with more and more frequently; besides this, the whole country was cut up by a network of dry aryks. The channel, throughout its whole length to Egen-Klych, bore the character of a large river with tolerably high banks. When there were sand banks in the middle of the bed, the channel of the river appeared along one or other of the banks. At Egen-Klych was found a dam about 124 sagens long. Near it were six wells containing but little water, in one of them only, fit for drinking, which placed the detachment in an extremely difficult position. Notwithstanding this, the Expedition was forced to bivouac at the wells of Egen-Klych.

The 3. of July, the Expedition moved on farther, also in two columns, with surveying parties. One column marched by a straight road to Ak-Bugut, and the other took the course of the old river bed. In the neighbourhood of Ak-Bugut there was a dam, and in all directions were seen dry canals and the remains of walls. At Ak-Bugut, for a distance of half a verst, there were 15 wells in two groups and there proved to be sufficient water for the whole detachment. The march was extremely heavy and tiring, the heat was intolerable, movement under the burning rays of the July sun was torture, shifting sand at times blocked the road by which the Expedition marched. Especially sombre news was received about the condition of the route ahead, joined with incredible difficulties and privations.

On the 4. of July, lay before the expedition the most difficult waterless march, through tracts containing bitter springs to the site of Dekcha. Already at Kunya-Urgench this march excited great fears. At Dekcha, according to the opinion of the natives, there might either prove to be no water at all, or it would be unfit for drinking, even for animals. The distance to Dekcha was not known exactly. It could easily prove, that only by the road it was about 70 versts, but by the bed of the Kunya-Darya, as the guides declared, the distance might easily reach 100 versts. After taking a supply of water, as far as this was possible, at daybreak, the Expedition marched out in two columns. The column with the train took the direction of Dekcha; and the column with the surveying parties followed the bed of the Kunya-Darya, and found the bed to a considerable degree filled up with shifting sand. Farther on, the bed turned into a deep ravine with high and steep banks. The guides affirmed that, having once entered that ravine, it would not be possible to ascend from it to Dekcha; that there was no water in the bed; that the distance by the bed was not known exactly, and that it was very probable that it would be necessary to spend the night in a waterless desert, without any provisions for man or beast. Impenetrably dense bushes, through which the column was forced to fight its way, a mass of shifting sand over nearly the whole extent of the route, several steep descents into the bed and steep ascents out of it; the great burden of spare water, and finally the burning rays of the sun and the intolerable heat tired in the extreme horses and men, who were forced to drag the waggons themselves or harness additional cavalry horses and themselves go on foot. Arrived at Dekcha, the bivouac was accompanied with every inconvenience. Both banks of the river bed here were perpendicular, and a descent to the bottom of the river was possible only in certain places. The bottom itself was covered with sheets of bitter water and overgrown with cane. The water in these pools had an extremely disagreable smell and taste, both from the presence of mineral salts and the organic remains decaying in it. The use of it hardly quenched thirst but only speedily deranged the stomach, both of men and horses.

From Kunya-Urgench, the Expedition set out under the predictions of inevitable destruction or the complete failure of the undertaking. Now the same was heard from the guides. They affirmed that on the site of Sary-Kamysh there was but one well, no forage, and that in the lake the water was disgustingly bitter, etc. Notwithstanding all these opinions, it was decided to go on farther.

The 5. of July, at three o'clock in the afternoon, the Expedition started towards the Lake Sary-Kamysh, the distance from which was supposed to be 35 versts.

The road was sandy and difficult. It was already dark, when it arrived at the bivouac. For this purpose, the bank of the old river bed was inconvenient; the descent to the well at the bottom of the bed was steep; water was scarce and an extreme want of it was felt.

The 6. of July, at daybreak, the Expedition undertook a reconnaissance under the cover of a sotnia of Cossacks.

Having left the wells about four versts behind, the Expedition perceived in the distance the blue surface of Lake Sary-Kamysh. But, to reach the sands surrounding it, it was necessary to cut one's way in Indian file through extraordinary thick and high bushes of tamarisk. The ground of the shore of the lake was very boggy and sticky. The air, fromthe great quantity of decayingorganic remains was filled with intolerable miasmas. The water of the lake was clean and clear, but gave forth an unpleasant smell, had a bitter taste, and was completely unfit to drink.

After having explored the banks of the Lake Sary-Kamysh as far as time permitted, the members of the Expedition turned back, and arrived at their bivouac at the wells of Sary-Kamysh about midnight.

The 7. of July, the Expediton left the wells of Sary-Kamysh and started on the road back, making supplementary surveys. Afterwards having made a reconnaissance away from the main road as far as the ruins of the fort Kizilcha-Kala, the expedition returned safe and sound to the town of Urgench. On the march from Kunya-Urgench to Fort Petro-Alexandrovsk the six sotnias of Orenburg Cossacks were appointed to serve in Turkestan. The Expedition also explored the Khan-Yab canal, to Yesaul-Bashi, Lauzan, Siuelli, and crossing to the right bank, proceeded to Fort Petro Alexandrovsk, carrying out surveys of the country on the way.

From the explorations made by the above named expedition, it appeared that, at present, the waters of the Amu-Darya do not flow throughout the whole length of the old bed of the Kunya-Darya. The dams artificially erected, as well in the bed of the Kunya-Darya as in the Lauzan, lead off the waters of the Amu-Darya in different directions from the old bed, the Kunya-Darya or the Urun-Darya, in consequence of which this bed remains completely dry, from the canal Shamrat to lake Sary-Kamysh and farther. But the dams existing in the old bed of the Urun-Darya, or the Kunya-Darya, as well as the numerous systems of dry aryks on both banks of the old bed, prove incontrovertibly that, not long ago, water flowed by the old bed of the Urun-Darya or the Kunya-Darya, and irrigated all the surrounding country. As a fact the natives affirmed that, 11 years ago, the water flowed as far as Llake Sary-Kamysh, and that the Lauzan was closed by a dam by order of the Khan of Khiva for the purpose of punishing the rebellious Turkomans on the banks of the old river bed, and forcing them to emigrate to the canals lying near to Khiva.

Thus, the different conjectures and hypotheses hitherto made about the existence of old beds were completely confirmed. It appeared that the old beds of the Amu-Darya (Darya-Lyk) and the Daudan not only actually exist, but that water flows by them over a considerable extent, that they are capable of taking water, which may be led into them, and that not long ago water flowed by the Kunya-Darya into the Lake of Sary-Kamysh.

The labours of the Expedition of the Urun-Darya, having proved the existence of the old beds and the recent flowing of the water of the Amu-Darya by the Kunya-Darya into lake Sary-Kamysh, excited the public interest and drew its peculiar attention to the farther exploration of the old beds of the Amu-Darya throughout their whole extent, for the more prompt solution of the question concerning the possibility of leading the waters of the Amu-Darya into the Caspian Sea. This circumstance impelled the Imperial Russian Geographical Society to send, in the year 1874, an Expedition for the purpose of examining this question as well as of exploring the Khivan Oasis and the neighbouring deserts. This Expedition, properly speaking, consisted of two divisions, which performed their operations independently of each other and pursued different objects. One of these parties, which included the zoologist Sievertsov, the botanist Smirnov, the meteorologist Dorandt, the ethnographer Sobolev, the hydrographer Zubov and Mr. Wood under the command of general Stolietov, passed partly from Kazalinsk trough Kizil-Kum, partly along the eastern shore of the Aral to the delta of the Amu-Darya and to Fort Petro-Alexandrovsk. But, as the administrative authorities of Turkestan did not wish to excite any fears or misunderstandings among the Khivans, this party was obliged to confine itself to explorations, on the right, Russian, bank of the Amu-Darya. The other part of the same expedition, consisting of Thillo, Solimani, Struve and Moshkov, made an exact survey of the levels between Kara Takmak, on the Aral Sea, and the Gulf of the Czesarevich (Miortvy-Kultuk) on the Caspian Sea, a distance of 346 versts. It then proved that the level of the Aral is 35 sagens higher than that of the Caspian Sea.

From the above cited sketch, it appears that the old bed of the Amu-Darya between the Khanate of Khiva and the Caspian Sea was explored from both ends, that is from the side of Khiva and from that of the Caspian Sea, and that only the middle part of it remained perfectly unknown. This omission was made good in the year 1875.

For the examination of the middle part of the bed of the Amu-Darya, from the wells of Igdy to lake Sary-Kamysh, Colonel Glukhovskoy organized a special caravan from Krasnovodsk to Khiva, and the authorities of the Caucasus sent the topographer Lupandin for the execution of surveying and topographical works.

The reconnoitring party, marching first, together with the detachment under the command of General Lomakin, to the wells of Igdy, marched farther under a feeble escort of Turkoman militia up the Uzboy, by the wells of Charyshly to Lake Sary-Kamysh. Thus, the topographer Lupandin made a survey of the Uzboy, from the well of Igdy to the wells of Dekcha, and joined the bases of the former, investigations, which were carried out, from the side of the Amu-Darya by Colonel Glukhovskoy, and from the side of the Caspian Sea by Colonel Stebnitsky.

In the year 1876, the Commission of the Urun-Darya under Colonel Petrusevich was organized from the side of the Caucasus by order of HIS IMPERIAL HIGHNESS, THE GRAND DUKE MIKHAEL NIKOLAEVICH.

This Commission took the levels of the country along the left bank of the river Amu-Darya from the town of New-Urgench almost to the town of Khodjeily, by the old bed of the Kunya-Darya from its exit at New-Urgench to the Lake Sary-Kamysh. Besides this, the levels of some canals were taken. By these observations, as well as by the transverse levels across the old river beds, the contour of the area explored was to some extent made clear. It appeared that the old beds, as well as the whole locality, have a common slope from the river Amu-Darya westwards, to the Lakes of Sary-Kamysh. Consequently, the results gained by the previous reconnaissances of the Urun-Darya Expedition in the year 1873 were completely confirmed. It was shewn that the passage of water out of the Amu-Darya by its old beds could not meet with serious obstacles. Also, the various theories on the geological upheaval of the soil as the cause of the interruption of the current of water by the old beds proved to be perfectly inadmissible. Together with this fact, was discovered the existence of the immense depression of Sary-Kamysh, the bottom of which is lower than the level of the Aral by 40, and even than the Caspian by 5.72 sagens. This depression, which lies in the line of the old beds of the Amu-Darya to the Caspian Sea, formed the bottom of the once existing Lake of Sary-Kamysh, which dried up completely after the interruption of the current of water by the old river beds.

During the high water of 1878, the waters of the Amu-Darya, having broken the dams which barred the old bed, the Kunya-Darya, ran by this bed and reached the Lakes of Sary-Kamysh.

In view of such an important event, Colonel Glukhovskoy when in the Transcaspian and the Caucasus, petitioned the Viceroy of the Caucasus to appoint a new expedition for the investigation of all the circumstances of the bursting of its banks and change of course of the Amu-Darya. HIS IMPERIAL HIGHNESS, THE GRAND DUKE MIKHAIL NIKOLAEVICH sent the engineer Helman to inspect the breaches, who took part in the operations of the former Urun-Darya Commission. His investigations demonstrated that the waters of the Amu-Darya, not finding room in their chief bed, as well as in the old beds and aryks, broke through the dams and the adjacent elevated bank and ruhed in the direction of the common inclination of the locality, drowning cultivated fields and villages. During the inundation, the waters of the Amu-Darya poured in such an mmense mass, that they submerged a considerable part of the Khivan Khanate. Even great towns, protected by strong walls, were not able to resist the force of the torrent and, only in a few places, the inhabitants, working to the number of several thousands on longitudinal dams, were able to save something from the raging river.

The greatest masses of the water of the Amu-Darya passed by he Kunya-Darya, the dams erected in this bed being broken through or passed round. Notwithstanding, however, all this abundance of water, running in during the inundation, the level of the Lakes of Sary-Kamysh did not rise more than 28 feet and, in consequence of this, there could not be any movement of water by the old channel beyond the lakes, the level of which lies considerably below the outlet of the Uzboy.

Comparing the results, gained by the Urun-Darya Commission in 1876 with, those obtained by the exploration of the inundation of 1878, it appears that, in the upper parts of the old channel where generally the water overflowed the banks evenly, the inclination of the high water approached very nearly that of the banks. Between Kunya-Urgench and the Egen-Klych dam, the fall of the high water diminished fairly regularly, thence beyond this dam and before the Lakes of Sary-Kamysh, the inclination increased suddenly, and became greater than that of the banks. Moreover, the dams exercised a considerable influence on the inclination of the high water. These, till their destruction, held up the water to a more or less extent. Up to the overflow of 1878, the old bed of the Amu-Darya was silted up with shifting sands, which in some places presented a continuous barrier of considerable dimensions. The water which broke through not only washed away in the shortest time sands accumulated by years, but also deepened the bed over an extent of 127 versts. To hat a degree the bed was cleared at this time, appears from the fact that, in winter even, a feeble current of water reached unhindered the Lakes of Sary-Kamysh. This fact, showing the force of the water, may serve as a refutation of the hypothesis that the change of the course of the Amu-Darya took place only from its silting up with sand.

CHAPTER II.

The organization of the Expedition formed, in 1879, by the order of HIS MAJESTY THE EMPEROR OF RUSSIA, for the exploration of the old bed of the river Amu-Darya, between the Sea of Aral and the Caspian. The execution of the operations in the delta of the Amu-Darya and in its old beds, between the Amu-Darya and the Caspian Sea.

In the year 1875, the waters of the Amu-Darya broke through the dams, moved along the Kunya-Darya and reached the Lake of Sary-Kamysh.

This fact, having proved the perfect possibility of the flowjug of the water of the Amu-Darya by the old bed of this river, the Minister of Ways of Communication, Adjutant General Possyet, in consideration of the extraordinary importance of this event, instantly put himself into relation with the authorities of the Caucasus and Tur kestan, as regards taking due measures for the solution of the question of a water way from Russia into Central Asia, which would be of first rate importance in all respects, both for Russia and for Central Asia. HIS IMPERIAL HIGHNESS, THE GRAND DUKE MIKHAIL NIKOLAEVICH, always interested in the ascertainment of the possibility of carrying Peter the Great's idea of the turning of the water of the Amu-Darya into the Caspian into effect, shewed great sympathy towards this enterprise. In the beginning of the year 1879, a Commission was organized, under the presidence of the Minister of Ways of Communication, composed of specialists and persons acquainted with the country, to consider the question of joining the Caspian with the Sea of Aral, by means of letting through the waters of the Amu-Darya into the Caspian.

This Commission, after having examined the then existing data upon this question, acknowledged the peculiar importance of a final solution of it, whatever the results of the explorations might be and found it necessary to organize for this purpose a special Expedition. consisting both of engineers and of representatives of various Ministries and Departments of the government and various branches of science.

It was decided to charge the Expedition with, the exploration of the Amu-Darya between the Sea of Aral and the Caspian, the investigation of this river, with the object of ascertaining the quantity of water. which it might be possible to direct into the old river bed upon the condition of preserving the cultivatiou of the Khivan Khanate. and the exploration of the delta of the Amu-Darya with the object of regulating it, in order to create a navigable way by the same, at the same time deflecting a part of the waters of the Amu-Darya into the old bed.

The 22-nd of May, the Emperor deigned to give His Imperial Assent to the opinion of the Council of State on the organization of the said expedition. It was placed under the command of Major-General of the General Staff A. Glukhovskoy, and was formed of the following persons: the Chief of the technical section of the Expedition, Holmstrem, the engineers Helman, Bolle, Svintsov, Balinsky, Maximovich and a corresponding number of engineers, topographers and officers of the Turkestan Military Topographical Section.

At first, it was proposed to immediately form and despatch parties of engineers to the places of operations, one to Krasnovodsk. and the other two to the section of the Amu-Darya. But the temporary commander of the Transcaspian division communicated that, in consequence of the disturbed state of the steppes, it was desirable to delay the despatch thither of the division. charged with the exploration of the old bed of the Amu-Darya. the Uzboy, to such time as it should be considered possible. In consequence of this, it was decided in 1879 only to send the party of the Amu-Darya, to explore the delta of this river, and the sending of the other two parties was deferred, it being proposed to form them in January of the next year 1880, if the condition of the steppes permitted it.

Labours of the Expedition in the delta of the river Amu-Darya.

A detailed programme was elaborated for the execution of the tasks of the Amu-Darya party, confirmed by the Minister of Ways of Communication. According to this programme, the party appointed to carry out the works in the delta was charged with the following labours: to make longitudinal and transverse levellings in the delta of the river; to observe the rapidity of the current of water at different points; to execute surveyss in various parts of the delta; to make observations on the level and the expenditure of water, and on the deposits carried into the Sea of Aral by the river itself and its tributaries; to explore the entrances to and exits from the arms; the properties of the shores; to perform other geological researches, and to collect various agricultural and statistical information.

For solving the problems, involved in the discovery of a navigable channel in the delta of the Amu-Darya, its regulation and finally the study of the question of establishing a navigable route be tween the Amu-Darya and the Caspian, the expedition carried out the following works.

Before all, it was necessary to ascertain whether it was possible, without harm to the economical condition of the Khivan Khanate, to lead off some of the water of the Amu-Darya into the Caspian Sea, and what amount.

For this purpose, in the autumn of the year 1879 hydrometrical stations were arranged at the following points: on the site of Tiuya-Muyun, at the entrance of the Amu-Darya into the Khivan Khanate, before the branching from it of the irrigation canals of the Khanate and of those of the Amu-Darya section. Another station was appointed opposite the separation of the river into arms in the delta and near the fortifications of Nukus, below the origin of the chief irrigation canals of the Khanate. Nenxt, stations were established at the points where the arms of the delta fall into the Sea of Aral, at the mouths of the Ulkun-Darya and the river Yany-Su; and besides these, there was a station in the delta of the Amu-Darya near the mountain Taz-Nura, for the purpose of observing the changes of level in the middle arm and the overflowings of the delta. At these stations, observations were made during a year, while sections of the running water were made, the velocity of the current was determined, soundings were taken and the quantity of drifts in the water was defined.

The engineering and topographical labours of the Amu-Darya party in the delta of the river began the 18. of October, 1879 from that point of the right bank of the Amu-Darya, opposite to which the explorations of the former Urun-Darya Commission of Colonel Petrusevich were terminated, that is about 4 versts below the outlet of the canal Suelli from the Amu-Darya. The party proceeded along the right bank of the Amu-Darya and along its middle arm Ishan, the overflowings and lakes of the central part of the delta and by the Ulkun-Darya, to the Sea of Aral.

The direction of the explorations by the middle arm of the Amu-Darya's delta was selected for the following reasons. 1. The western arm of the delta, the Taldyk, is barred by dams and at its fall into the sea is silted up with mud, in consequence of which navigation by it is impossible. 2. The eastern arm of the delta of the Amu-Darya, consisting of the Kuvansh-Djarma and the Yany-Su, presents great difficulties to navigation, and at the same time was formerly at different times more or less explored. 3. By the middle arm of the delta (the Ishan, the lakes and the Ulkun-Darya) no explorations had ever before been undertaken, and only in the navigation season of 1879 for the first time, the «Samarkand», a steamer of the Aral flotilla, with barges, made a successful voyage. As the middle arm of the delta flows through a locality covered with lakes, floods and extensive marshes, and is for the most part inaccessible for works from spring to autumn, that is during high water in the Amu-Darya, it was necessary to choose for the purpose the months of October, November and December, the time of low water in the Amu-Darya. The labours in the delta of the Amu-Darya consisted of longitudinal levellings, measurement of the bed and determinations of the velocity of the tream and expenditure of water.

The Ishan for the first 30 versts flows tolerably regularly. its banks during low water are solid enough to permit the execution of works, but afterwards diminishing in height they gradually became weak, then tenacious and swampy and finally disappear entirely under water where the yare only defined by dense growths of cane, while the stream of the channel is entirely free from it. From the thirtieth verst onwards, the Ishan presents great difficulties in the way of the execution of any kinds of works on its banks. Having brought the operations to this point, it became necessary to choose one of two courses: either to continue the works in winter, taking advantage of the ice, or defer them till spring and carry them out by making a circuit of several versts at a distance from the chief arm.

It was decided to continue the works, notwithstanding the frosts which reached 20° R., and by means of the ice to carry them on as far as the solid ground, whence they might be continued at all times of the year. From the thirty-fifth verst from the source of the Ishan, the works were conducted on the ice. The levelling was greatly impeded by setting up the instruments on a frozen crust, while on the other hand the same circumstance facilitated movements. By means of the ice it was possible to reach the isolated lakes, completely inaccessible at the warm season of the year. At the same time, it is much easier and more convenient to cut a road through the dry cane than the green.

Thus, the works on the ice were pushed on a distance of forty versts as far as the Kushkan-Tau hill, whence there were no difficulties in continuing them farther in the spring, as although floods are met with below this hill, they are of considerably less extent. Towards the end of January, the ice had become so unsafe, that it was dangerous to continue working upon it. Therefore, the winter labours of the Amu-Darya party were closed the 30-th January 1880 at a spot on firm ground on the Taz-Nura hill, two versts from Kushkan-Tau. The Amu-Darya party returned to fort Nukus and occupied itself arranging the results of the field operations. In March 1880, after the arms of the delta of the Amu-Darya had become free from ice, a special Commission under the presidentship of the Chief of the technical part of the Expedition, the Engineer Holmstrem executed a general reconnaissance of these arms, which lasted from the 15-th of March to the 16-th of April 1880. The exploration extended to: 1. the eastern arm and in particular, the streams Kuvansh-Djarma, Purkhan and the river Yany-Su; 2. the middle arm, consisting of the stream Ishan, lakes and the river Ulkun-Darya; 3. the western arm, composed of the stream Bi-Djab, the Kunya-Darya and the Taldyk, and also the part of the Kunya-Darya between the Taldyk and the river Ulkun-Darya.

After having made the exploration of the delta of the Amu-Darya, the Commission proceeded to the examination of all the circumstances of the undertaking and of all the facts collected.

Having compared with each other the arms of the Amu-Darya delta in respect of the degree, in which each of them wanted regulation, in making choice of the direction for a navigable way into the Sea of Aral, in case of turning the waters of this river into the Caspian, the Commission at a general meeting on the 26. of April came to the conclusion, that the middle arm, the Ishan, the lakes and the Ulkun-Darya present the best conditions for the most convenient establishment of navigation, best of all secure the further existence of the said navigable way, and require at the same time the least expenditure of money.

In consequence of this conclusion of the exploring Commission, it was decided to continue the works of the Amu-Darya party on the middle arm, i. e. on the Ishan, the lakes and the Ulkun-Darya, to the Sea of Aral. Thus the 9. of May 1880, this party recommenced its works from the station erected by it in winter on the Taz-Nura hill. The spring labours also presented difficulties of their own. It was exceedingly difficult to cut clearings through the fresh, green cane. The bank on which the levelling had to be executed in many places was distant several versts from the chief arm, which there was no possibility of approaching with the level.

From June the hot weather set in, and then there appeared myriads of gnats and midges. Extensive floods compelled the widening at this point of the survey to as much as eight versts. Peculiar difficulties in executing the operations continued up to the beginning of the Ulkun-Darya which, after receiving the waters from the floods, flows in the form of a regular river into the Sea of Aral. For this extent, the obstacles and inconveniences in the execution of operations were no longger met with as in the floods of the Ishan.

Besides these operations in respect of the exploration of the middle arm of the delta, the Amu-Darya party executed a junction of the levellings of the Urun-Darya Commission and the Imperial Expedition, between the forts of Nukus and Kunya-Urgench, and also a junotion of the hydrometrical post Tiuya-Muyun with the permanent station of the Urun-Darya Commission.

After the termination of the Amu-Darya party's operations on the central arm of the delta, the explorations of other arms and channels of the delta were continued in the same autumn, 1880; the levelling on the eastern arm Kuvansh-Djarma being carried out as far as the Daukarian lakes. On the western arm, surveys were made on the Kunya-Darya. Besides this, the levelling between Thillo Station and the level of the Sea of Aral shewed a falling of the level of this sea amounting to about one arshin in six and a half years.

Supplementary surveying operations in the Amu-Darya delta were continued partly as late as 1881 and 1882.

Thus, the following operations were executed in the delta of the Amu-Darya:

In 1879, 1880.

1) Double levelling, 391 versts	782 versts.
2) Single »	289 »
Total.	1071 versts.
3) Instrumental survey on a scale of 0.01 sagen to 100 sagens	634 square versts.
4) Half-instrumental survey, 0.01 sagen to 2 versts	5600 » »
Total.	6234 square versts.

In 1881.

1) Single levelling	150 versts.
2) Instrumental survey on a scale of 0.01 sagen to 100 sagens	251 square versts.

In 1882:

1) Instrumental survey on a scale of 0.01 sagen to 100 sagens	135 square versts.
2) Half-instrumental, 0.01 sagen to 2 versts .	2068 » »
Total.	2203 square versts.

Thus, in the delta of the Amu-Darya was executed a total of:
Levellings . . . . , ... 1221 versts.
Surveyings. . . . . 8688 square versts.

LABOURS OF THE EXPEDITION IN INVESTIGATING THE ANCIENT BEDS OF THE AMU-DARYA.

In 1880, the Imperial sanction was given for the expedition against the Akhal-Teke and therefore it was proposed at first to send only one engineering party to Krasnovodsk to carry out investigations on the side of the Caspian, taking advantage of the march into the steppe of the troops of the expedition; while the formation of the Amu-Darya party was put off until the state of things in the steppes was cleared up.

Soon, however, in consequenee of the inconvenience of detaching from the Akhal-Teke expedition a specially strong escort for the work of the party, in the direction of the Caspian Sea, it was decided to put off the organization and despatch of the party to Krasnovodsk.

Only in the direction of Khiva, in the autumn of 1880, was authorisation given to transfer a part of the Amu-Darya party to the Khivan steppes to carry out operations, if that appear possible, taking advantage of the movements in the steppes of the Sary-Kamysh active detachment, stationed by order of the Emperor for the securing of the frontiers of the Khivan Khanate from the attack of the Teke Turkomans and to cover the furnishing of General Skobelev's detachment with those supplies which it may prove possible to derive from the Khivan Khanate. To attain these objects, and to cut off the communication between the Teke's and the Khivan Khanate and the Khivan Turkomans, the Sary-Kamysh detachment stationed till the end of August on the Kunya-Darya, behind the town of Kunya-Urgench, near the Balykly wells, was transferred to the southwestern frontier of the Khanate near the ruins of the fortress of Izmukshir. Taking advantage of the movements of this detachment, an investigation of the steppes and ancient channels was instiuted. For this purpose, in the middle of September, an engineering party consisting of two engineers and two topographers, to which was afterwards added a geologist and two more engineers, arrived at Fort Izmukshir on the western frontier of the Khanate. Hence to the Uzboy extends a vast sandy desert in which, for a distance of from 225 to 250 versts, there was then no well with water fit for use, all having been filled up with sand.

The operations of the Expedition were to be carried out in that part of the Aralo-Caspian depression, which is included between the Amu-Darya, the steppes of Merv, the steppes of Akhal-Teke, the Caspian, the mountains of Balkhan and the plateau of Ust-Urt, at the foot of which is situated the immense depression of Sary-Kamysh about a hundred versts from the frontier of the Khivan Khanate. The levellings of the Urun-Darya Commission showed, that the level of the lakes of Sary-Kamysh is more than 40 sagens below that of the Sea of Aral and 5.72 sagens below that of the Caspian. Thus, to reach the Caspian, the waters of the Amu-Darya must first fill up this basin, which would need a great quantity of water and much time. Accordingly, to solve this question satisfactorily, the Expedition was obliged to carry out investigations and levellings in the basin of the Sary-Kamysh and, in the Khivan steppes to the east of this basin, to find the direction, in which it would be possible to let the waters of the Amu-Darya through into the Caspian Sea avoiding the basin of Sary-Kamysh. Gigantic difficulties and obstacles presented themselves to the accomplishment of these works. The whole vast steppe region of the Expedition's operations, from the western frontier of the Khivan Khanate to the Caspian, for a distance of more than 800 versts, consists of poorly watered or waterless desert steppes, marching through which was considered not only difficult but even impossible for a body of troops. Not only to stay for any length of time, for work in these steppes, was considered impossible, but even small caravans and single riders rarely dared to cross them. The little knowledge possessed about them increased still further the difficulties and obstacles. In summer, the steppes assume a dead aspect; all vegetation is parched up, the heat in the hot season exceeds 40^0 R.; the water in the wells becomes very bad and quite unfit to drink. Thirst, during the heat is intolerable, and water is found only in a few wells and is mostly of a bad quality. These steppes were to such an extent impassable that, during the Khivan Expedition in 1879, Colonel Markazov's detachment could not reach Khiva and was forced to retreat, after only going as for as the wells of Igdy; and colonel Skobelev, having made a reconnaissance of the steppe from Fort Izmukshir to the wells of Orta-Kuy in 1873, confirmed the fact of the impossibility for troops to traverse the steppes.

In consequence of such difficulties, when forming the plan of campaign in the beginning of 1880 for general Skobelev's Akhal-Teke Expedition, it was acknowledged to be impossible to march the Turkestan troops immediately through the steppes. But the Expedition hoped to be able to execute the operations with which it was charged, and to accomplish the march through the waterless steppes by taking the following indispensable measures. It was determined to choose autumn, and even winter, for explorations and movements over the steppes, as the most convenient seasons of the year, to make every effort to properly equip and protect the Expedition; to form a good camel-train; to provide a sufficient supply of all kinds of provisions and especially of water-raising apparatus, without which no march could be thought of. In such a state of the steppes, it was impossible to march over them, and it was absolutely necessary first of all to furnish the route with water by digging wells, for which it was necessary to move on gradually, carrying water 50 and even 100 versts.

The digging of wells proved to be very difficult. After a well was dug, there proved to be little water in it, or it was not fit for use, and it was necessary to dig a new one. The soil, in which the wells had to be dug, often consisted of loose sand which, notwithstanding the casing made of boards, often fell in and, on one occasion, the men who were digging a well were buried by the sand falling in and were with difficulty saved. Notwithstanding these difficulties in furnishing the route with water and of marching through poorly watered steppes, by the middle of October, the wells were dug from Izmukshir to Fort Giaur-Kala, for about 125 versts surveys and reconnaissances of the localities were executed, and a description of the discovery of the ancient canal Cherven-Yab was made. Next, the organization of a levelling party was taken in hand. But, in consequence of the disturbed state of the steppes, over which bands of Teke Turkomans were moving, once even a Kossack post at Kizilcha-Kuyusy being attacked by them, it became necessary to push on a part of the detachment to the wells of Charyshly, whence the bands commonly made their raids, and then begin the levelling. In the beginning of November, the detachment occupied these wells without meeting any bands and, at the same time, a telegram was received from the Commander of the Turkestan Military District ordering the whole Sary-Kamysh detachment to return to Petro-Alexandrovsk and to afford every assistance to the detachment of Colonel Kuropatkin, then marching towards the Oasis of Akhal-Teke, to effect a junction with the troops of Adjutant-General Skobelev for united action with the Caucasian troops against Geok-Tepe.

The detachment of Colonel Kuropatkin arrived at the Chagyl wells without water-raising apparatus and sufficient transport and forage, and the Sary-Kamysh detachment, which was covering the Expedition, detaching a company of sharpshooters and experienced men of the rocket platoon, gave them to the Turkestan detachment, as was done with its water-raising apparatus, its best transport camels and some stores, in consequence of which the Expedition was made completely incapable not only of accomplishing its chief labours, but even of making necessary reconnaissances. It was only possible to send one engineer to make a reconnaissance of the steppes from Giaur-Kala, near the end of the former Chermen-Yab canal, to the wells of Orta-Kuyu and farther by the old bed of the Amu-Darya to the wells of Igdy.

After that, the operations of the Expedition for the year 1880 were concluded, and it returned together with the Sary-Kamysh detachment to fort Petro-Alexandrovsk. It was made clear by these reconnaissances, executed in the autumn of the year 1880, that there was much more work to be done to explore the steppes and the old river beds than was anticipated. It was necessary to make further supplementary engineering explorations, on the one hand of the Daudan from its commencement to the Sary-Kamysh Lakes and from the latter, by the old river bed to the Charyshly wells and, on the other, by the Chermen-Yab canal and through the steppes, also to the Charyshly wells, in order to solve the question of the possibility of getting round the Sary-Kamysh depression.

The works of the Expedition, in the year 1880, gave the following results:

1. The desert, dead and waterless steppes between Khiva, Orta-Kuyu, the Uzboy and Charyshly, which were considered as impas-

sable for troops were, for the first time, traversed by a detachment of Russian troops, explored, and furnished with wells at the most important points. They were found practicable in all directions, particularly in the autumn, as the Expedition supposed before, when drawing up the plan of operations.

2. The reconnaissances and surveys, executed in the Khivan steppes, cleared up the question of the importance of the old beds of the Amu-Darya and showed that a water way might be established from the Amu-Darya by the Daudan to the site of Chagyl, for a distance of 150 versts, in consequence of which the distance by land between Chagyl and Kizil-Arvat, then the terminus of the Transcaspian railway, might be reduced to 460 versts.

3. The march from Izmukshir to the Charyshly wells shewed the existence there of the ancient Chermen-Yab canal, about 111 versts in extent, along which formerly lay the chief trade route from Khiva to Turkomania and Persia.

A series of ruins of towns and forts, Dauda-Kala, Airtam, Kizilcha-Kala, Shakh-Senem, Giaur-Kala, and traces of former settlements shew that these localities were inhabited, tilled and flourishing. Here was found even an old river bed, which the natives called Tonu-Darya (Kanga-Darya), winding round the elevation of Kanga-Kyr. All these facts shewed that there exists a slope of that locality in that direction from the Amu-Darya to the Uzboy, and that there is a possibility of letting the waters of the Amu-Darya flow to the Charyshly wells around the Sary-Kamysh basin even, if the Daudan should prove inconvenient for this purpose.

4. The geologist, Prince Gedroitz, made a geological examination of the steppes from the western frontier of the Khanate through Kizilcha-Kala, Giaur-Kala and Orta-Kuyu to the wells of Igdy, and the engineers made a technical inspection of the Turkoman steppes, the roads across them, the old river bed of the Uzboy, between the Igdy wells and those of Bala-Ishem, and of the locality between the Charyshly wells and Orta-Kuyu. Thus, the reconnaissances and technical inspection of the steppes shewed that the most important and difficult operations lay before the Expedition, on the space from the end of the Daudan and the Chermen-Yab to the wells of Orta-Kuyu and that, upon the complete and detailed exploration of this part, depends the successful solution of

the question of the passage of the waters of the Amu-Darya to the Caspian.

The explorations of the Khivan steppes, carried out in the autumn of 1880, were of great service to the Expedition, because it was then determined, in which direction the levelling and surveying operations must be accomplished by the Expedition in the steppes of the western part of the Khivan Khanate. The explanation of these questions formed at that time the chief and essential problem of the Expedition, as determining the operations lying before it in 1881. Besides this, it became clear that it was necessary to do much more work in the steppes than was anticipated in forming the original plan of operations, namely more than 450 versts in length, as it proved to be necessary to execute the levelling not only of the dried up part of the Daudan, adjacent to the steppes, but also of the part of it which is in the limits of the Khanate, as well as of the Chermen-Yab canal.

The labours of the Expedition, during the autum of 1880, shewed, among other things that, to secure success to steppe operations, it is necessary for each party to have not less than five men for the execution of the engineering and three men for that of the topographical works, in consequence of the severe local conditions, which had a disastrous effect on the health of the workers. In the autumn of 1880, nearly all the engineers fell ill, and both the military topographers accompanying the Expedition were so bad that they were unable to make reconnaissances.

On the basis of the reconnaissances and surveys, made in the Khivan steppes in 1880, the plan of operations for 1881 was formed and sanctioned, according to which it was determined to execute the works by means of two engineer parties placed on the side of the Khivan Khanate in order, after bringing to a conclusion the operations in the Turkestan Military District, as far as the Charyshly wells, to continue them from these wells along the Uzboy in the Caucasian Military District. For the works from Khiva to the wells of Charyshly, working parties were appointed from the Amu-Darya troops. In the autumn, for the continuation of the works from the wells of Charyshly, an escort was to come from the Caucasian troops to relieve the Turkestan working parties.

The 12-th of February, the expedition and its escort, consisting of two companies of infantry, a sotnia of Cossacks and a rocket platoon, started From fort Petro-Alexandrovsk in the direction of the site Uaz and, the 28-th of February, reached Toprak-Kala. But up to the middle of March, there was no possibility not only of carrying on operations, but even of verifying levellings, in consequence of continuous violent snow storms, winds and bad weather. As for carrying on operations by means of two parties, it was neccessary for three engineers to come from St. Petersburgh and three topographers from Tashkent. The first steppe party was organized before their arrival. It consisted of four engineers, and three military topographers of the Turkestan Military Topographical Section.

Operations were begun, the 17-th of March 1881, from Colonel Petrusevich's Urun-Darya Commission station, near to the ruins of Fort Mirish-Kala, thence by the Daudan and farther by its dried up bed to the former Lake of Tinniu-Kliu, in the region of the Sary-Kamysh basin, then by the Tonu-Darya and further through the steppes to the Charyshly wells. After pushing the operations to the Charyshly wells, on the 17-th June, the first steppe party was transfered to the floods of the Daudan at the bridge on the road from Tashauz to the village Ak-Saray; thence operations were directed by the Daudan and Shakh-Abat to the Amu-Darya. The engineering labours of the first party were concluded on the 15-th of August and the surveying works, the 20-th of that month.

On the arrival from St. Petersburgh of three engineers, and of three military topographers from Tashkent, recently ordered to join the expedition, the second or Engineer Svintsov's steppe party was organized, consisting of four engineers and two topographers, while two topographers were sent to work in the delta of the Amu-Darya. The second steppe party began its operations, the 2-nd of June 1881, from the fort of Mangyt-Kala along the Daudan, first by its dried up bed and then by its floods in the Khivan Khanate. The party carried its operations to the beginning of the works of the first party and terminated the engineering operations the 15-th of August, and the surveys the 24-th of that month. In proportion as the works were finished, the officers of both parties returned with their men to the Petro-Alexandrovsk fort for the purpose of furnishing themselves with all that was necessary for the contemplated operations in the steppes in the autumn.

The 4-th of September 1881, the Expedition, consisting of the two steppe parties, again marched out into the steppes. For the operations, two working companies were appointed from the troops of the Amu-Karya division, each a hundred and twenty men strong.

The first party proceeded, the 14-th of September 1881, to work on the ancient Chermen-Yab canal, from where it leaves the Daudan, near to the site of Chagyl, and directed its operations by the canal to the high ground of Kanga-Kyr, and by its north eastern slope to the wells of Charyshly. The second party began to work, the 24 of September 1881, from the Sary-Kamysh springs, joined here the works of the Amu-Darya Commission, explored the Sary-Kamysh basin to the Charyshly wells, where, it arrived at the end of October. By this time, the Caucasian escort, consisting of a company of His Imperial Highness the Grand-Duk, Alexander Mikhailovich's Crimean Regiment and of 30 Kuban Cossacks, also came to the Charyshly wells. The 1-st of November, the Turkestan detachment was sent back to Petro-Alexandrovsk and, for the works of the second steppe party, from the Charyshly wells a working party was assigned from the Caucasian escort. On the arrival of the first engineer party at the Ide-Khauz pools, about 30 versts short of the Charyshly wells, the second Turkestan working company was sent back to Petro-Alexandrovsk, and replaced by a Caucasian one.

Meanwhile, in November 1881, when the native guides confirmed the view that, near to the elevation of Zengi-Baba there, is a river bed by which, in past times the water of the Amu-Darya used to flow from Chardjui, a special party was organized for the purpose of clearing up and verifying the information. This party, consisting of a corps of military topographers, one officer and an escort of six Cossacks under the command of an ensign, was ordered to reconnoitre the Chardjui bed from Zengi-Baba to the Shiikh wells, and thence to the Uzboy. The 20-th of November, the party marched out of camp, and started for the Laili and Shiikh wells, and thence westwards to the Bala-Ishem wells, where it arrived the 2-nd of December 1881. Thus, during 13 days, a reconnaissance was made for a distance of 220 versts at the most severe season of the year, when snow had fallen, frosts and fogs had begun, and severe blizzards were raging.

In general, the operations of the Expedition in the steppes in

1881 were accompanied by extraordinary difficulties and all kinds of privations and it demanded a great exertion of strength and energy, to surmount all the obstacles presented by works conducted in waterless deserts, at first during the extreme heats and afterwards winter cold. The Sary-Kamysh depression was especially distinguished by the badness of the water. During the works here, it was sometimes necessary to bring water from a distance of more than 70 versts. At the same time, it was precisely in the Sary-Kamysh depression that the Expedition was obliged to execute the most difficult and detailed works, to determine the dimensions and form of the basin, and to choose the direction of the canal to avoid it. For that purpose, transverse sections were made, reaching 57 versts in length and, during the works along the western slope of the Kanga-Kyr elevation and of the Ide-Khauz depression, it was necessary every three versts to make transvere sections 8 versts in length, so as to obtain, in the direction chosen for the canal avoiding the Sary-Kamysh basin, levelling points every three versts. Moreover, 32 versts of longitudinal and transverse levellings were made in the hollow, extending southwards from the Ide-Khauz depression, between the sites of Ak-Navat and Hok-Navat.

This manner of executing the works greatly retarded their advance while, at the end of November, severe frosts and snowstorms set in. In December, the frosts reached—19° R. The men warmed themselves only at camp fires. Water froze in the barrels, which burst even with iron hoops. In the camp of the engineer party, there was no water during several days, although an entirely sufficient provision of barrels with frozen water was at hand. It became necessary to stop the transport of water, and the parties found themselves obliged to get water for drinking or cooking from snow or ice. There was little snow, and to gather and melt it was a matter of great difficulty. On account of the frosts, snowstorms and strong winds, the operations went on slowly and often stopped altogether. The forage for the camels became very scarce, because the steppes were covered with snow; the camels, accordingly, began to grow weak. Under such conditions, it was not possible to carry out any more levelling operations without detriment to their exactness; therefore, the second steppe party terminated its labours, the 6-th of December 1881, at the site Siradja, and the first party carried its works by the 11-th of December as far as the wells of Charyshly. After this, the members of both parties executed reconnaissances of the whole locality situated to the east of the Uzboy, between the wells of Ikedje-Kuyu to the north, and the wells of Bala-Ishem and Kurtysh to the south, for the purpose of ascertaining whether there was not any other river bed falling into the Uzboy from that side; but nothing of the kind was found, and the reconnaissances were terminated the 23-rd of December 1881.

In all, during the years 1880, 188 the Expedition executed in the Turkoman and Khivan steppes the following operations:

1. Double levelling for a distance of 570 versts . .	1140 versts.
2. Single longitudinal and transverse levelling . .	2022 »
Total	3162 versts.

3. Instrumental survey on a scale of 0.01 sag. to 200 sag.	2106 squ. versts
4. Half instrumental survey, on a scale of 0.01 sag. to 1 verst.	9815 » »
5. Reconnoitring survey, on a scale of 0.01 sag. to 5 versts.	7953 » »
Total	19874 squ. versts.

The continuous, prolonged and exhausting operations, executed by the same members of the expedition, first in the delta of the Amu-Darya in the years 1879, 1880, and then in the steppes in the autumn of 1880 and during the year 1881, under the most difficult climatic and local conditions, impaired the health of the men to such a degree, that those engaged in the work were constantly ailing, and their farther remaining in the steppes might have entailed serious and dangerous illness.

There was no longer any possibility of continuing the operations in the steppes. It became necessary to give the members of the expedition some repose. The replacing of the former members by new would entail much more considerable expense and, at the same time, the new comers, acquainted neither with the local peculiarities and features of the Amu-Darya and its old beds nor with the operations already performed, would not be competent to carry out the programme, with which the expedition was charged.

Besides, such a mass of results of field works, of materials and facts, was already accumulated that it became necessary to arrange them with the help of the same persons, who had been engaged in carrying them out, so as to make use of the results already gained for future operations. In consequence of all these circumstances, permission was given, to carry out operations in 1882, only in the Khivan Khanate and in the district of the Amu-Darya, by means of a special party, while the remaining members of the Expedition were allowed to return to St. Petersburgh to arrange and elaborate the materials gathered in the previous years.

Accordingly, a special topographical party was organized, consisting of a corps of military topographers and two engineers, for making plans in the Khivan Khanate and partly in the delta of the Amu-Darya, and carrying on hydrometrical works on the Khivan canals for the purpose of calculating the quantity of water expended in irrigation, of examining the methods of irrigation in practice in the Khanate, and of gathering statistical information concerning the Khivan Khanate and its communications.

The 2. of January 1882, Captain Kosanchich started from the wells of Bala-Ishem with a detachment of 26 Cossacks and arrived, the 18. of January, at Petro-Alexandrovsk, while the Expedition with the Caucasian escort, having started from the wells of Igdy the 3, of January arrived in safety at Fort Kizil-Arvat, the 11. of January, 1882.

The 24. of January, 1882, the topographical party set out with the intention of accomplishing its task in the Khivan Khanate and, during the year 1882, executed surveys on the canals Siuelli, Yakub-Yargan, Lauzan, Nurak, Saubet-Jargan, Mangyt-Arna, Shamrat and Khazavat, as well as of the locality between the Shamrat and the Kunya-Darya and between the canals Yarmysh and Klych-Niaz-Bai, also completing surveys between Kungrad and the southern coast of the Sea of Aral and the eastern branches of the Ust-Urt, and on a part of the western arm of the Bi-Djab.

Besides these operations, hydrometrical works were conducted along the chief canals of the Khivan Khanate, the system of irrigation was explored, the quantity of water expended in irrigation was estimated, and various statistical data were collected upon the Khivan Khanate.

Besides hydrometrical works on the canals, the following operations were carried out in the Khivan Khanate in 1882:

1) Instrumental surveys, on a scale of 0.01 sag. to 100 versts	88 squ. versts.
2) Half instrumental surveys on a scale of 0.01 sag. to 1 verst	4106 » »
3) Reconnoitring surveys on a scale of 0.01 sag. to 2 versts	1414 » »
Total	5608 squ. versts.

In the year 1883, the Expedition had to terminate the operations on the Uzboy from the site of Siradja, the place where the operations were stopped in 1881, near to the wells of Gech-Geldy, to the Gulf of Balkhan, a distance of about 450 versts. According to the plan of operation sanctioned, it was intended to execute levelling and surveying works in the spring, in April and May, and in the autumn, in August, September, October and November. In spring, it was intended to begin the operations from the wells of Bugurudja or Alty-Kuyu, and to direct them by the Uzboy to Balkhan Bay, so as to work in April and May not far from inhabited places, near the line of the Transcaspian railway. The Expedition intended to spend the months of June and July in Krasnovodsk, arranging the results obtained by the operations executed on the Uzboy in the spring because, during the summer, in the Transcaspian desert sandy steppes the temperature rises to 49° R., rendering it quite impossible to execute any work in the open air. For the autumn operations, the Expedition intended to start from Kizil-Arvat, the 25-th of July, in the direction of the wells of Igdy and thence to direct its operations by the Uzboy to the wells of Kurtysh, Bala-Ishem and farther to the site of Siradja. After terminating its labours there, the Expedition was to move on to the wells of Igdy and, late in the autumn when less water is required, and when it is easier to preserve, to execute works on the side of the Uzboy, presenting difficulties from the absence of water, from the wells of Igdy to the spot whence the spring works of the Expedition began. To accomplish all these operations, the members of the expedition must arrive from St. Petersburgh and from Fort Petro-Alexandrovsk before the 20-th of March at the Transcaspian railway station of Bala-Ishem, so as to begin the field work on the 1-st of April.

In accordance with the plan sketched out, the expedition arrived at the Transcaspian railway station of Bala-Ishem from St. Petersburgh at the end of March. At the same time, a working party arrived consisting of 55 privates of the 25-th Transcaspian Battalion of Sharpshooters under the command of Ensign Strzalkowski and 25 Cossacks of the Taman Regiment under the command of Ensign Kiashko. The 26-th of March, the Expedition started from the station of Bala-Ishem in the direction of Lake Topiatan whence, on the 2-d of April, levelling operations by the Uzboy were commenced in the direction of the Balkhan Gulf. The topographical party started from Fort Petro-Alexandrovsk, the 28-th of February, 1883, with an escort of 8 Cossacks and, having made a march through the steppes, arrived at Fort Kizil-Arvat, the 5-th of April, and at the station of Bala-Ishem the 7-th of the same month. The Cossack escort then returned to Fort Petro-Alexandrovsk by the same route. During the operations of the Expedition on the Uzboy, near the wells of Dzoyuruk, a meteorogical station was founded, the 5-th of April; it was afterwards transferred to Fort Kizil-Arvat, where observations were begun from the 6-th of July. The operations of the spring period were carried by the Expedition to Balkhan Bay, the 21-th of June, 1883, and the members of the expedition were ordered to go to Krasnovodsk where, during the month of July, they occupied themselves arranging the results obtained from the works in the steppes. During the stay of the Expedition in Krasnovodsk, the heat was intolerable, it rose to 39°R. in the shade. At night blew a stifling wind; the men fell into a state of complete exhaustion, and nearly all the members of the Expedition were ill. But, notwithstanding this, on the 31-th of July, the Expedition started for the autumn operations from Fort Kizil-Arvat to the Uzboy, namely to the wells of Igdy. This march was extremely difficult and even dangerous. An incredible heat still continued in the steppes, the road lay over hills of shifting sands, and the water in the wells of Naurly was so bad, that it was entirely unfit for drinking, even affecting the health of the horses and camels. Men, horses and even camels began to suffer from indigestion and diarrhoea. During the march from the wells of Naurly to the wells of Igdy a distance of about 45 versts, men, horses and camels arrived at a state of complete prostration. During this march, fourteen sheep and two dogs accompanying the working party fell from exhaustion in the sands. Although the Expedition reached the wells of Igdy, it was enfeebled to an extreme degree.

The 8. of August, the Expedition began its levelling and surveying operations from the wells of Lower Igdy up the Uzboy to the site of Siradja, where they were terminated the 21. of September. Then, the Expedition moved to the wells of Igdy and, having begun operations there the 29. of September, directed them down the Usboy. At the same time, Engineer Svintsov was charged with the definite solution of the question of the Chardjui river bed, by taking levels from the site of Kugunek on the Uzboy to the wells of Islam-Kuyu and farther along the Unguz. Having finished this levelling, a distance of 73 versts, and having explored the locality as far as the wells of Shiikh, Svintsov returned on the 17. of October to the wells of Igdy and joined there the engineering party.

The results of the levelling, and the topographical character of the region between the Uzboy and the wells of Shiikh, a distance of about 100 versts, as well as the absence of shells and muddy river deposits, prove that the Unguz represents a number of separate hollows and ravines, the lowest points of which have a fall from the Uzboy in the direction of the wells of Shiikh. In consequence of this fact, Unguz cannot be allowed for this distance to be an old bed of the Amu-Darya, the water of which could not flow from the lower lying Unguz into the Uzboy.

The 9. of November 1883, the levelling operations of the Expedition were carried as far as Topiatan and terminated at the station erected at the commencement of the work of the spring period. The surveys continued to the middle of December. Thus, at the end of the year 1883, the field works relative to the exploration of the old bed of the Amu-Darya, the Uzboy, were brought to a close.

The works of the Expedition in the year 1883 were carried on in desert steppes remote from inhabited places, and were accompanied by immense difficulties in a hot climate amidst sandy deserts. The heat in the daytime and the stifling air at night began already in the month of May. The bed of the Usboy throughout the greater part of its length, from Lake Topiatan to the Balkhan Gulf, contains bitter water. The depth of the water is in general insignificant. The measurements were therefore made by fording the river and, in places where the water was very deep, by means of rafts built of barrels. Movement along the quaking and sticky banks of the river bed and in its bottom, either by fording or on rafts, presented great difficulties. The feet sank deep in the salt mud, and the dry crust on the surface, as it broke, wounded them. During the exhausting operations of the spring, accompanied by great privations, the Transcaspian railway served as a great support and comfort. By it water was supplied, which rendered it possible at least to quench the thirst and to forget to some extent days spent amidst privations and suffering, especially during the operations down the Uzboy from the bitter wells of Alty-Kuyu. In the autumn, when the Expedition moved on into the depths of the steppes by the Uzboy between the site of Siradja and Lake Topiatan, to the former difficulties was added the necessity of limiting all wants to a minimum, called forth by the remoteness of this country from any inhabited places and by the total absence of water between the wells of Igdy and those of Tagalek, where for a distance of nearly 100 versts there is not a well containing drinkable water. Water had to be transported from the two extreme points, that is over a distance of 50 versts.

From the absence of water, this is the most difficult part of the Uzboy. Works could here be executed with success only in the autumn, when the water flowing by the Uzboy was nearly everywhere covered, in consequence of a very extensive evaporation, with an icy crust of salt and gypsum and, in consequence, no longer presented such impediments to a passage across the bed as it did in the spring.

The following works were executed during the year 1883 in the Turkoman steppes:

1) Double levelling 558 versts	1116	versts.
2) Single longitudinal and transverse levellings	$1062^1/_2$	»
Total	$2178^1/_2$	versts.

3) Instrumental survey on a scale of 0.01 sagen to 200 sagens	1723 square versts.
4) Ditto, on a scale of 0.01 sagen to 50 sagens	10 » »
5) Half instrumental survey on a scale of 0.01 sagen per verst.	1546 » »
6) Route survey on a scale of 0.01 sagen to 4 versts	3115 » »
7) Ditto, 0.01 sagen to 5 versts . . .	1450 » »
Total	7844 square versts.

Taken all together, during the Expeditions execution of works concerning the exploration of the Amu-Darya old beds, in the Khivan Khanate and Transcaspian steppes, the quantity of works accomplished was as follows:

1) Levelling works.

a) Double longitudinal levelling of 1128 versts	2256 versts.
b) Single levelling	$3084^{1}/_{2}$ »
Total.	$5340^{1}/_{2}$ versts.

2) Topographical works.

a) Instrumental surveying on a scale of 200 sagens and less to 0.01 sagen	3927 square versts.
b) Half instrumental surveying on a scale of 1 verst to 0,01 sagen	15467 » »
c) Reconnoitring on a scale of 4 and 5 versts to 0.01 sagen.	13932 » »
Total	33326 square versts.

After the termination of the field operations, in the beginning of February, 1884, the Expedition returned to St. Petersburgh and proceeded to the arrangement of materials and facts, gathered during its field occupations, and those from the former Urun-Darya Commission, which in 1876 and 1877 executed levellings along the old river bed, the Kunya-Darya, from the Amu-Darya to the Sary-Kamysh Lake, along the Amu-Darya from the mouth of the Kunya-Darya to the mouth of the Suelli canal, and along the Lauzan, SaubetYargan, Shamrat, Mangyt-Arna and Biuten-Yab canals. The elaboration and arrangement of the materials obtained by the former Urun-Darya Commission presented great difficulties, in consequence of the death of its Chief, Colonel Petrusevich, and the absence of its members.

CHAPTER III.

Results of the operations, executed by the Expedition in the delta of the Amu-Darya along its three chief arms: a) the Eastern arm, consisting of the Kuvansh-Djarma channel, the Daukarian Lakes and the river Yany-Su; b) along the Middle arm, consisting of the Ishan channel, the lakes, and the Ulkun-Darya; c) the Western arm, consisting of the Bi-Djab channel, the Kunya-Darya and the Taldyk. Scheme for the regulation of the Middle arm, chosen for the establishment of water communication between the Amu-Darya and the Sea of Aral.

The hydrometrical operations executed, during the years 1879 and 1880, at the stations erected on the site of Tiuya-Muyun, near Fort Nukus, at the mouths of the Ulkun-Darya and the Yany-Su and at the foot of the mountain Taz-Nur, for the determination of the quantity of water, which could be carried off from the Amu-Darya without any detriment to the tillage of the Khivan Khanate, as well as for the determination of the quantity of deposit, carried by this river, etc., demonstrated that the quantity of mud carried by the waters of the Amu-Darya varies, at different seasons of the year; the smallest quantity of sediment is carried in December, January and February at low water, the greatest quantity in May, June and July, at flood level. Relative to the quantity of sediment, the Amu-Darya is one of the muddiest rivers; although less so than the Nile and Ganges, it surpasses in this respect many other rivers.

The results of the hydrometrical operations of the Expedition, at these above mentioned stations, are represented by the following table:

1879 — 1880.	At Tiuya-Muyun	At Fort Nukus.	Irrigation of the Khivan Oasis.	Falls into the Aral.	Lost in the floods
CUBIC SAGENS PER SECOND.					
The greatest expenditure of water	435	345	90	220	125
The least expenditure of water	85	70—76	10—15	40	30—35
The average expenditure of water	190	150	40	115	35
Per cent . .	100%	—	21%	60½%	18½%

From this table it appears that, out of 190 cubic sagens of water entering the Khivan Khanate on an average for the year per second, only 115 cubic sagens, or 60½ per cent, reach the Aral; 40 cubic sagens, or 21 per cent, are expended on the irrigation of the Khanate; 35 cubic sagens, or 18½ per cent, are lost in the floods of the delta, the maximum expenditure of water being 125 cubic sagens, and the minimum, 30—35 cubic sagens.

This quantity is lost in evaporation and filtration through the ground in the floods and lakes, and only a part of it is expended on the irrigation. This water in the lower parts of the Amu-Darya overflowing the banks in summer time, inundates the crops and villages and causes great disaster to the inhabitants of the delta, seriously hindering the development of agriculture.

Taking into consideration that the coasts of the Sea of Aral are extremely deserted and lifeless, and that the preservation of its present level is not absolutely necessary to the prosperity of the surrounding country, the more, as the once existing steam navigation on the Syr-Darya is done away with, and the waters of that river are employed more and more on the irrigation of the neighbouring regions, it proves to be necessary only to regulate one arm of the delta to establish steam communication by the Amu-Darya with the Sea of Aral, and to leave to the remaining arms only the quantity of water needed for the irrigation by the adjacent fields. As to the quantity of water which thus remains to be disposed of, it was decided that it might be used, by sending it along the old bed of the Amu-Darya into the Caspian.

The area of the cultivated land in the delta of the Amu-Darya at present amounts to about 800 square versts, 400 of which are situated on the left Khivan river side, but, in view of the general development of the cultivation of the country and the gradual increase of the fixed population, it was assumed that the area of cultivated land, the irrigation of which would have to be secured in any case, would amount to 1000 square versts, for which, taking also into consideration the consumption of water by the inhabitants, filtration through the ground, evaporation, the careless management of water used in irrigation and other unproductive expenditure, 8.85 cubic sagens per second, on an average, would be necessary. For the regulation of the middle arm of the Amu-Darya delta, with the present expenditure of water, 28.3 cubic sagens will be absolutely necessary; this makes in all 37.15 cubic sagens of water per second on an average for the whole year.

From the above table, it appears that at Fort Nukus pass and enter the delta of the Amu-Darya on an average 150 cubic sagens per second; deducting from this 37.15 cubic sagens, needed for the regulation of the middle arm of the delta and for irrigation, there results 112.85 cubic sagens of water per second, which could be sent by the old bed of the Amu-Darya without any detriment to the cultivation either of the Khivan Khanate or of the delta of the Amu-Darya, keeping up communication at the same time between the Amu-Darya and the Sea of Aral [1]).

The Amu-Darya river, about 200 versts before it falls into the Sea of Aral, makes a turn near to Fort Nukus from a northwestern direction nearly to west, and in a course of about 45 versts divides into three arms, Eastern, Middle and Western, which carry the waters of the river into the Aral, and several channels which fall into the lakes existing iu the central part of the delta The vast area of land bounded by the two extreme arms, the Eastern and Western, the main river and the Aral, containing about 9,000 square versts, forms the delta of the Amu-Darya, consisting chiefly of lakes, floods and marshes nearly completely overgrown by impenetrable cane. On this vast surface, overflowed in various degrees by water, more than usual at high water, possessing a very fertile soil, consisting of mud and humus, dwells an insignificant Karakolpak population, of about 14,000 families, which is employed on small and less frequently flooded patches of ground, about 400 square versts altogether, between the canals Kegeili and Tarly, and on the western arm of the delta near the town of Kungrad.

In winter time the delta becomes more lively. Besides its permanent Karakolpak population, many Kirgiz are seen, who come

---

[1]) Such a putting of the question does not at all decide the quantity of fields in the Amu-Darya's delta capable of cultivation; it only solves theoretically the question of the maximum quantity of water, which may be carried of from the Amu-Darya without any detriment to the existing cultivation of the Khivan Khanate and the Amu-Darya delta.

If, to allow the waters of the Amu-Darya to pass into the Caspian and to irrigate cultivable fields in the Turkestan steppes, less water would be necessary, than what is above calculated (112.85 cubic sagens per second) as proves really to be the case, of course nothing would prevent an increase of the surface of fields to be cultivated in the delta of the Amu-Darya. In reality, the quantity of cultivated fields in the delta of the Amu-Darya, taking into consideration the diminution of the floods and the falling of the level of the Sea of Aral, might easily reach twice as many square versts, the irrigation of which, as will below be seen, would perfectly harmonize with the letting through of the waters of the Amu-Darya into the Caspian, as well as with the irrigation of the Turkestan steppes, and is conditioned by the carrying off of a part of the Amu-Darya water from the delta of that river, where it at present only harms and hinders the development of agriculture.

there with all their herds from the Mangyshlak, Orenburg and even the Ural steppes, where frost and blizzards are very severe, and deep snow hides the forage from the cattle, while in the delta no such frosts exist and the cane presents a rather coarse but good forage, as well as excellent fuel.

But, in summer, myriads of gadflies and gnats appear in the delta. Besides this, on the cane and thorns a certain kind of beetle occurs which, according to the opinion of the natives, when swallowed by the camels, causes their death. For this reason, in the beginning of the spring, the Kirgiz quit the delta with all their herds and, taking from Kungrad goods intended for transport to Orenburg, at first take the direction of the western coast of the Sea of Aral and then across the steppes, driving their herds from one pasturage to another, the whole steppe being at that time of the year covered with excellent forage.

In the beginning of May, the last Kirgiz retire from the delta and then, up to the month of October, only the Karakolpak population remains there. In the delta, there are no considerably populated places, the Karakolpaks live in small auls near to their fields, and there is a bazaar only on the bank of the Kegeili canal at Chimbai, where the neighbouring population meet twice a week to buy or sell what they need; for the rest of the time, only a few dozen families employed in industries live in Chimbai. The whole delta of the Amu-Darya forms the district of Chimbai, the chief of which has his residence in the little Russian Fort Nukus, situated on the right bank of the river Kuvansh-Djarma. At present, the delta of the Amu-Darya is in the following condition:

The Easten arm leaves the Amu-Darya two versts below Fort Nukus, forming the Kuvansh-Djarma and the Kok-Uziak channel. The Kok-Uziak is rapid and tortuous, and has a width of 20 to 70 sagens. Its right bank is accompanied by a dam. The Kok-Uziak, 3 versts from its mouth, falls into the Kuvansh-Djarma.

At the 11-th verst from the mouths of the Kuvansh-Djarma, the Tarly canal leaves it on the left which, together with the Kegeili, somewhat below it, belongs to the most ancient irrigating canals of the Amu-Darya delta.Below the confluence of the Kok-Uziak, on the right bank of the Kuvansh-Djarma, sand hills extend, which do not disappear as far as the Daukarian Lakes. The Kuvansh-

Djarma, taking a north easterly direction from the Amu-Darya, 5½ versts below the confluence of the Kok-Uziak, makes a short turn to the south, detaching to the left the Kegeili. This canal has very considerable dimensions and flows in the direction of the town of Chimbai. Below this town, the Kegeili flows to the north-west in the direction of the elevation of Kush-Kana-Tau and falls into lakes, while its other arm, known under the name of Nu-Pyr, flows first to the north-east and then to the north.

After having detached the Kegeili canal, the Kuvansh-Djarma narrows and, running on about a verst, turns again to the north-east. About 10 versts from the outlet of the Kegeili, a channel called Djalpak leaves the Kuvansh-Djarma on the left; this channel falls into a vast depression between Kegeili and the Kuvansh-Djarma, which serves as a basin for the superfluous waters of these channels. From the exit of Kegeili, fields are cultivated on both banks of the Kuvansh-Djarma, in small patches or narrow strips of land, beyond which are to be seen floods overgrown with cane, and sand dunes.

The channel Kru-Uziak leaves the left bank of the Kuvansh-Djarma 15 versts below the Djalpak, and takes a north-easterly direction. Some 50 versts more, below the Kru-Uziak, the Purkhan channel separates from the Kuvansh-Djarma on the left and falls into the western of the Daukarian Lakes, the Kara-Teren. Farther on, the Kuvansh-Djarma, making short bends and detaching the channels Bais and Tampi, continues its course under the name of Kara-Kol to the north-east, and falls into the eastern of the Daukarian Lakes, the Kungrad-Kul. The length of the Kuvansh-Djarma is 130 versts. Its breadth fluctuates between 30 and 115 sagens. Its banks, at first high, become lower and lower so that, during an insignificant rise, the river overflows its banks. The depth of the Kuvansh-Djarma is fairly uniform and reaches for a considerable distance seven feet.

In proportion as the river approaches the Daukarian Lakes, its depth diminishes, permanent floods overgrown with cane appear, and the current becomes weak. The bottom is mostly covered with sand and occasionally with small stones; in the lower part, it consists of clay with sand.

The velocity of the current of the Kuvansh-Djarma is, in the low water season about 3, in the high water about 5½ feet per second. This river is very tortuous, and many of its bends are sharp and short. After the fall of the water, the clayey and sandy banks fall in great blocks, which are washed away by the stream and then settle and form sand banks. Their formation is helped by the roots of trees and bushes, which fall into the water together with the underwashed banks and, in high water time, float till they meet the first obstacle. The mud carried by the water in great quantities then settles over them, forming first sand banks and afterwards small islands.

There are a great many of these sand banks at the beginning of the course of the Kuvansh-Djarma, and they present a considerable obstacle to steamers entering it from the Amu-Darya. In high water time, these sand banks are covered with water, but in time of low water they reappear.

The Daukarian Lakes consist, properly speaking, of two, situated near each other and joined by channels. The north-western lake, the larger, is called Kara-Teren, and the south-eastern, Kungrad-Kul. These lakes, in some parts, have their banks overgrown with cane, amidst which bays and open sheets ot water appear, bearing among the natives names as lakes, separate and independent of one another. The area of the Daukarian Lakes is calculated to be, during low water, about 258 square versts and, at the high water season, about 327. On the north and east, these lakes are bounded by a clayey salt strip of land, bordered by the Bel-Tau-Hills, consisting of a clayey sandstone and covered by crystalline gypsum. These hills advance close to the northern part of lake Kara-Teren and, forming its shores, bear various names.

To the north of these hills, a salt and sandy steppe begins, called Kizil-Kum, which extends to the banks of the Syr-Darya. Saltmarshes spread to the east of the lakes and floods, to the south and west. The latter are covered with cane, and supplied with water by the Kuvansh-Djarma. The avarage depth of lake Kara-Teren and its northern bay of Djiuriuk is about 3.7 sagens. The winds cause such an agitation of the surface of the lakes, that navigation on them in the local boats becomes nearly impossible so that, to facilitate the passage of these kayuks from the Kuvansh-Kjarma into the river Yany-Su, they are sent by the Purkhan and by the Uyezd-Yargan channel, and thus avoid the roughness of the water of the Daukarian Lakes. In the neighbourhood of the Daukarian Lakes, the Kirgiz of the Chimbai section of the Amu-Darya district lead a nomadic life; they are engaged in cattle breeding, and charcoal burning from saksaul, which overgrows in considerable masses the eastern and northern parts of the lakes, as well as the banks of the river Yany-Su.

The Yany-Su carries the waters of the Daukarian Lakes into the Sea of Aral. The weak current of the Daukarian Lakes grows somewhat stronger in the south-western part of Lake Kara-Teren. About 6¹/₂ versts from the island Monatai where, from both sides, sandy hillocks approach close to the lakes, a channel is formed. The current grows stronger there, and it is from this point that the rise of the river Yany-Su is reckoned, although it is accompanied still somewhat farther by floods and lakes. At first, the Yany-Su flows between sandy hills, and the river supplies many floods with water. The clear water of the Daukarian Lakes remains the same in the Yany-Su, which has a very irregular course and makes frequent and sharp bends. The banks of the river are for the most part low, in many places they scarcely rise above the high water level; the banks are sandy, and here and there hillocks, covered with saksaul bushes approach and retreat from the river. The marshes are overgrown with cane and, in the lower parts of the river, tamarisk bushes are to be found. Down the river, 34 versts from its source, the Yany-Su flows through a ridge of sandstone; this point of the river is known by the natives under the name of Tash-Utkul, which means «stone ford», and is now a ford. Five versts below this, an arm called Ike-Ata detaches from the Yany-Ku, on the left; this branch falls into the Sea of Aral, and 19 versts below the Yany-Su falls into the Lake Kly, which on the north joins a bay of the Aral. Ike-Ata, in consequence of its shallowness, especially at its outlet, is inaccesible even for local boats. Not far from Ike-Ata cultivated fields, land sown with melons and cucumbers, and Karakolpak auls are to be met with on the banks of the Yany-Su, covered with bushes. Below the exit of the Ike-Ata, here and there amidst the floods, are to be seen sand hillocks and rice fields. The nearer to the sea, the lower the banks are, and consequently are oftener inundated, vast marshes being formed. Navigation on the Yany-Su is considered possible, although the shallowness on the bar in Lake Kly and the crookedness of the Yany-Su present great difficulties and obstacles.

The middle arm of the delta quits the Amu-Darya 33 versts from Fort Nukus, and consists of the Ishan about 70 versts in length, lakes about 30, and finally of the Ulkun-Darya 75 versts it length, the character of each of these parts sharply differing from that of the rest. This arm carries into the Aral the greatest quantity of water of the Amu-Darya which, for a distance of 33 versts, from Nukus to the exit of the Ishan, presents a completely regular and normal river, and can be navigated even by vessels drawing about ten feet of water.

The Ishan, from the quantity of water flowing into it out of the Amu-Darya, represents so to say a direct continuation of this river, and hence the supply of water rose every year, its banks were continually washed away, and its dimensions augmented, to the disadvantage of other arms and affluents.

The general characteristic of the Ishan is that, possessing weak muddy, and sandy banks, it has a tolerably considerable fall, and a velocity which reaches 9 sagens, in consequence of which the banks are washed and carried away. The Ishan often branches and, for a considerable part of its course, flows through great lakes, which do not dry up during the whole year and are overgrown with dense cane and shrubs. These, on decaying, form a manure, which contributes to the growth of the soil, still further increased by the muddy deposits of the river during the summer high water.

After leaving the Amu-Darya, the Ishan trends for about 15 versts south and north, then turns to the northwest, flowing in that direction also about 15 versts, when it makes a turn to the northeast and, after having so flowed about 20 versts, resumes its former direction. During the said course, it frequently separates into branches, which afterwards join again into one arm.

At a distance of 41 versts from the outlet of the Ishan, an affluent called Tilla-Bai falls into its left arm; this affluent separates from the Amu-Darya not much below the Ishan. Some 54 versts from where it leaves the Amu-Darya, the Ishan falls into the lakes.

The banks of the Ishan, as far as the town of Kron-Tau, consist of mud and sand, mixed with mould produced from cane. From the point mentioned, they are clayey. The Ishan, getting gradually wider, forms the lakes of Savan-Kul and Kos-Kul, and falls into

Lake Ai-Kul, which has a surface of about 12 square versts. The western part of this lake is full of sandbanks, the eastern is tolerably deep and bordered by cane bushes; while the southern part is shallow. Kushkana-Kul with its floods forms a continuation of Lake Ak-Kul which is, in its turn, a prolongation of Lake Ai-Kul. Lake Kushkana-Kul occupies an area of about 8 versts, and obtained its name from the hills Kushkana-Tau, situated on its eastern shore. Farther come the lakes Kara-Kul, Djaman-Kul, At-Irgen, Bir-Kazanly-Kul, Little Sary-Kul and Great Sary-Kul. All these lakes have the same character; amidst the marshes and the cane, bordering them, are to be seen sheets of open water, of different sizes and shut in by cane. The shores are everywhere low and covered with cane. The channel is irregular, approaching at one time one shore at another the other. It goes through narrow passages from one sheet to another, it being often necessary to avoid islands and cane growths. The floods of the middle part of the delta extend westwards to the hills of Kube-Tau and Tumalak-Tau, consisting of clay, sand and ferruginous sandstone. The area of all the floods of the central part of the delta may be reckoned at least 1,100 square versts, the lakes properly speaking covering about 200 square versts with an average depth of 10.5 feet. In consequence of the great quantity of cane on the western and eastern sides, this basin must in course of time become narrower; the cane, rotting and fallsing, gets covered over with muddy sand and clay, especially at highwater.

Therefore, the areas of the floods must gradually diminish and also the quantity of water evaporating. The effect of the force of the current must contribute to the formation of river beds. So, at present, besides the formation of deposits in the cane, a gradual although slow formation of permanent banks out of the deposits is to be observed, marking out the limits of the bed of the Ishan amidst the floods and lakes.

Out of the north western part of the lake, water flows by two channels at about a hundred sagens from each other. The western is called Kuk-Darya, and the eastern, Shkly; both carry their water into the Ulkun-Darya. The total navigation from the junction of the Ishan with the Amu-Darya to the Ulkun-Darya is 105 versts.

The Ulkun-Darya, from the confluence of the Kuk-Darya, over

the whole distance of about 77 versts to the Sea of Aral, now presents a rare specimen of a remarkably regular river.

The Ulkun-Darya, as regards its depth and width, possesses a particularly regular and majestic appearance for the first 63 versts before its branching into arms, not far from its fall into the Aral. The Ulkun-Darya at present carries into this sea directly nearly the whole quantity of water, which it receives from the floods by means of the channels Kuk-Darya and Shkly. This river, 6.5 versts below the ruins of Ak-Kala, parts into two chief arms, by which its waters reach the Sea of Aral. The right arm, called Ulu-Ayage, has a northern direction, and falls into the Aral 12 versts from the branching of the river. The left arm, called Kichkine - Darya, from its northwestern direction represents, so to say, a prolongation of the Ulkun-Darya, and falls into the Aral 14 versts after separating from the main river. The dimensions of the transverse section of the Kichkine - Darya are also uniform, the width is 50 to 100 sagens, the depth in most cases 3 sagens.

The Ulkun-Darya, in its present state, has the appearance of a fine navigable river from every natural point of view, and requires no outlay for its regulation.

It is open to navigation even by vessels of considerable draught, with the exception of the entrance from the Aral, where there is a bar only admitting the passage of steamers with flat bottoms. In recent time the steamers of the Aral Flotilla entered the Ulkun-Darya by the Kichkine-Darya arm.

The high banks of the Ulkun-Darya, below the point where the Kuk-Darya falls into it, render it impossible for the inhabitants of the delta to engage in agriculture, because for this purpose it would be necessary to raise water by means of «chigirs» or water wheels, which method of irrigation is generally speaking not practised by nomad tribes. From the present state of the Ulkun-Darya's level, the natives wandering on its banks conclude that the supply of water into the Ulkun-Darya is gradually diminishing. But in reality it does not seem to be so, because there are no fords now as there were in former times.

Ruins of settlements and of fortified places as also abandoned cultivated fields, to be met with on the banks of the Ulkun-Darya, testify that in former times they were populated and that the population was a settled one.

Both banks of the Ulkun-Darya are covered with vast pastures offering excellent feed for cattle, intermingled with thinly growing cane. This is why the Karakolpaks and Kirgiz, engaged in cattle breeding, wander over these regions.

The western arm of the Amu-Darya delta leaves it 10 versts below the outlet of the Ishan, and consists of the Bi-Djab channel 44.5 versts in lenght, of the Kunya-Darya up to the dam Ulu-Bugut, 32 versts, and of the Taldyk 75 versts. Its total length is 151.5 versts.

In 1884, a continuous stream along the western arm of the Amu-Darya delta existed only as far as the town of Kungrad; along the Taldyk, the water did not reach the Sea of Aral, because it is dammed up, and the whole insignificant quantity of water, which it receives, is consumed in the irrigation of the neighbouring fields. The part of the Kunya-Darya between the town of Kungrad and the Ulkun-Darya is dammed up for the purpose of raising the water level, so as to enable it to be led on to the fields requiring irrigation. The water, carried round the dams by means of small canals called «salmas» only reaches the lowest of the former, the Ilmrat Bugut.

At present the Bi-Djab, separating from the Amu-Darya in a northerly direction, after having traversed a distance of 10 versts, makes a turn to the northwest, and then regains its northern direction, turning slightly westwards, and near its end, after running about 41 versts, making a sharp turn to the west, retains this direction to the pointwhere it falls into the Kunya-Darya.

On both banks of the Bi-Djab some versts below its exit from the Amu-Darya, in consequence of the low banks, lakes and flooded lands overgrown with cane are formed. They are fed with water by the Bi-Djab at the season of high water in the latter.

On the right side these floods begin, where the channel out of the Tilla-Bai falls into the Bi-Djab, and extend along the bank of the Bi-Djab as far as Oguz, a channel flowing in a northwestern direction to the lakes and floods, into which the Ishan falls. East, these floods extend to Tilla Bai and the Ishan, and, on the left bank of the Bi-Djab, the floods begin from the channel Mamy and extend over a strip of land on the bank of the Bi-Djab 3 to 5 versts in width, and approach the Kunya-Darya where Sarkrauk falls into it. These banks must be drowned in the high water season, and the water flowing over them feeds the floods and leaves its deposit on them ever, more and more filling up hollows and places, overgrown with cane. Thus, during the high level season, only a part of the water, flowing out of the Amu-Darya into the Bi-Djab, reaches the bed of the Kunya-Darya and by it the town of Kungrad. The greater part is lost in the floods, extending on both sides of the Bi-Djab. From this, fact it results that, without artificial help, the water of the Bi-Djab would not reach the Kunya-Darya, even in case of an increase of the current of water out of the main river into the Bi-Djab.

The Bi-Djab, as it has been said, after a course of 45 verts, falls into the bed of the Kunya-Darya. This bed, from the place where the Bi-Djab falls into it, represents a continuation of the former bed of the Amu-Darya which, below the exit of the Bi-Djab, is filled up with sand up to the bank's edge, so that it is no more visible. Farther on, some traces of it are yet to be seen.

The bed of the Kunya-Darya, after the junction with the Bi-Djab, represents a ravine bounded by scarcely elevated banks, the right bank being lower than the left one. Into this ravine the waters of the Bi-Djab fall and, as a narrow streak, traverse a distance of about 32 versts, up to the dam of Ulu-Bugut, erected in the bed of the Kunya-Darya, near the town of Kungrad. This bed of the former river has for the most part a nortwestern trend. It is crooked and is 40 to 90 sagens in breadth and is so far silted up, that the banks of the former river rise obove the level of the water flowing in the bed not more than 0.2 to 0.4 sagens. The bed of the Kunya-Darya which, from the exit of the Taldyk to the Ulkun-Darya, might be called the Kunya-Ulkun-Darya, after the outlet of the Taldyk, turns nearly at a right angle to the east and has a breadth of 28 sagens. Its bottom is higher by one sagen than that of the Taldyk. In this place, there is standing water in the form of a lake, which extends to the dam Esaul-Bashi; farther come the dams, Kiat-Bugut, Taz-Bugut, Kazak-Bugut, Ishan-Bugut, Hodja-Bugut and Ilmrat-Bugut. Nine versts below the dam Ilmrat-Bugut, the Kuk-Darya falls into the bed of the Kunya-Darya; hence the Kunya-Darya obtains the name of Ulkun-Darya and, in its farther course to the Aral, is like a continuation of the middle arm of the Amu-Darya delta. Ovet this distance, the left bank of the Kunya-Darya is generally more elevated than the right. The soil of the right bank is mostly sandy clayey and overgrowen with small bushes; it is not rare on this bank to meet with salt marshes bare of all vegetation and covered with a crust of salt. The soil of the left bank consists of rotten cane and mud and is overgrown with cane. Cultivation is very insignificant on the banks of the Kunya-Darya, and is principally siauated on the left bank. The population is devoted chiefly to cattle raising; only a small part to agriculture.

A number of aryks or ditches have been cut through the fields for carrying water from the Kunya-Darya, but a great many of them are in a neglected state. The cultivated fields are situated either on the very banks of the river or at a short distance only from the water line, when the space between the river bed and the fields is covered with shrubs.

The Taldyk, from where it leaves the Kunya-Darya to the Irniaz-Bugut dam has a breadth beginning with about 30 sagens and, just before the dam, about 70 sagens. It flows almost in a northwestern direction, between steep banks. Farther on the water overflows, to the whole breadth of the river bed, and thus reaches the dam of Irniaz-Bugut. There is a dam, Kiat-Bugut, 13 versts lower down. About 32 versts below Kiat-Bugut, the Taldyk separates into two arms; the left retains the name Taldyk, while the right is called Medeli, in its turn also branching into two arms. The right arm, flowing eastwards, retains the name of Medeli and is barred by a dam of the same name; the left arm, flowing in a northeastern direction, is called Kasyk-Uziak and is barred by two dams called Kasyk-Bent. The main bed of the Taldyk, after the branching from it of the Medeli, is also barred by a dam bearing the same name.

The actual banks of the Taldyk consist of a solid clayey and sandy soil. A part of the Taldyk, namely that which is close to the dam of Kiat-Bugut, has a tolerably regular appearance, and the nature of the soil on its banks is identical with that of the surrounding locality, completely covered with rotten cane, mud and sand. The water level in the Taldyk before the dam Kiat-Bugut, is considerably higher than that in the lower part of it. The population of the banks of this part of the Taldyk, taking advantage of the pressure of the water are consequently enabled easily to lead water by aryks into their fields and, accordingly, here are engaged in tillage. Above the dam Kiat-Bugut all vegetation soon disappears. The dams existing in the Taldyk were erected for the purpose of raising the level of the bed for the irrigation of the fields.

The vast lakes and floods of the middle part of the Amu-Darya delta, formed chiefly by the waters of the Ishan arm, further receive water from the channels of the Eastern and Western arms of the delta as well as the channels, which have an independent exit out of the Amu-Darya above and below that of the Ishan. Among the most considerable channels of the part of the delta in question are the Chartombai, Tiulegen and Tillia-Bai, issuing from the Amu-Darya and the channel Oguz, from the Bi-Djab.

From a general exploration of the delta, the Expedition came to the conclusion that the middle arm,—the Ishan channel, the lakes of the middle part of the delta and the Ulkun-Darya—best fulfil the conditions for the establishment of a navigation between the Amu-Darya and the Aral.

The central arm in its present condition is, during the midsummer water-level, in both the Ishan and lakes and in the Ulkun-Darya, on the whole, satisfactory for the navigation of river boats with a draught of not more than four feet.

Navigation may meet with difficulties only in a few places in consequence of the narrowness and tortuousness of the channel. In these places works must be executed, consisting in the widening of the bed, the straightening of its windings and the regulation of the entrance of the Ishan into Lake Ai-Kul, where a bar is forming which may, in consequence of the deposit of sediment here, present difficulties in the future to the passage of steamers. These works will require an insignificant outlay. Seeing that, if a part of the Amu-Darya's waters were directed by its old bed into the Caspian, the quantity of water in the river during the midsummer water-level would diminish, due measures must be taken, so that the same quantity of water approximatively may flow into the middle arm of the delta as flows at present into the Ishan during midsummer water-level, and with a depth sufficient to navigate river boats. Starting from this point of view, the regulation of the central arm of the Amu-Darya delta must consist in such a distribution of the mass of water entering the delta, that it may be preserved in the now existing arms and channels in the quantity, which is needed for irrigation and the other wants of the population, and that out of the remainder such a mass may be concentrated in the middle arm, as is necessary for creating a navigable way out of the Amu-Darya into the Sea of Aral. Consequently, for the regulation of the middle arm, works must be executed, on the Amu-Darya from the Nukus hydrometrical station to the outlet of the Ishan, a distance of 33 versts; along the Ishan from the point where it leaves the Amu-Darya to its fall into the lakes, and through the lakes up to the fall of the waters of the central part of the delta into the Ulkun-Darya. According to the scheme of the Expedition, all the above mentioned works for the regulation of the central part of the delta could be executed in two years; their cost, according to the estimate made, is determined to be about 500,000 roubles.

For the purpose of comparing this estimate with the cost of the establishment of navigation by the western arm of the delta, the Bi-Djab, Kunya-Darya and Ulkun-Darya,— a scheme for a route by this arm was drafted. The cost of the works to be executed was calculated at two and a half million roubles, in consequence of the unfitness for navigation of the present bed of the Bi-Djab, instead of which it proves to be necessary to dig a new canal with a dam of great dimensions on its right bank, so as to prevent its waters from breaking through into the lakes of the central part of the delta, towards which a considerable fall exists there.

As to the eastern arm of the delta,—the Kuvansh-Djarma, the Daukarian lakes and the river Yany-Su,— it proved to be impossible to establish navigation by it in consequence of its extreme shallowness, crookedness and the absence of solid ground on the banks, especially in the lower part of the course of the Yany-Su.

CHAPTER IV.

Results of the works, executed along the old beds of the Amu-Darya: the Kunya-Darya, the Daudan, the Sary-Kamysh basin and the Uzboy. Schemes for letting the waters of the Amu-Darya through into the Caspian: 1) by the bed of the Kunya-Darya, the Sary-Kamysh basin and the Uzboy; 2) by the Lauzan canal system, the Kunya-Darya, the Shamrat canal, the bed of the Daudan, the canal avoiding the Sary-Kamysh basin and the Uzboy; 3) by the Shakh-Abat canal, the bed of the Daudan, the canal avoiding the Sary-Kamysh basin and the Uzboy; 4) by the bed of the Kunya-Darya, the canal Shamrat, the bed of the Daudan, the canal avoiding the Sary-Kamysh basin and the Uzboy.

THE Amu-Darya after leaving the mountains, before entering the limits of the inhabited part of the Khivan Khanate, flows along a valley, whose bottom presents so to say a succession of hollows or basins joined with each other by the river bed. This valley has steep banks, sometimes diverging for many tens of versts from each other, at others approaching each other to a few hundred sagens. Below the site of Tiuya-Muyun, the banks of the Amu-Darya valley diverge to the right and left, and the river traverses an immense plain, which slopes in a northwesterly and westerly direction.

Eastwards from the Amu-Darya, this plain is covered with shifting sands, with the exception of an insignificant patch of land on the right bank of the river which is fit to live on. The region westwards from the Amu-Darya presents, over the whole distance with few exceptions to Lake Sary-Kamysh, an even surface and a soil quite fit for cultivation. This enormous plain is bounded on the east by shifting sands, out of which southwards from the Khivan Khanate elevations project occasionally, which are seen on approaching the Sary-Kamysh basin. Westwards, the Sary-Kamysh basin is bound by the plateau of Ust-Urt and the Sea of Aral. The sands on this plain are concentrated eastwards from the Amu-Darya, chiefly in its old bed, the Daudan, as far as to where it crosses the canal of Shamrat, and also on the eastern border of the Sary-Kamysh basin.

Below the site of Tiuya-Muyun, the two banks of the Amu-Darya have an entirely different and peculiar appearance. The left bank forms the far stretching perfectly flat Khivan Oasis, covered with luxuriant fields and numerous gardens, in the shade of which latter are scattered the clay huts of the population. There also extend vast canals, bordered on both sides by trees. Along the right bank on the contrary, with inconsiderable interruptions, lie sterile dune-like sand hills in rows, invading this plain from the north and northeast, from the Kizil-Kum. The narrow strip of land varying in width between the river and the hills, is in some places settled and cultivated and, near the water for a short distance, partly bordered with dense growths of turangil and djida. Beyond this zone of bushes, which sometimes does not exceed in width fifty sagens, extends a plain covered with sparse grass, and isolated thorn bushes.

Throughout the whole of its length, from the site of Tiuya-Muyun to the Sea of Aral, the Amu-Darya only comes across one ridge in its course, namely that of Sheikh-Djeili which seems to be cut in two by the Amu-Darya. Its branches also pass over to the other, that is the left side of the river; naked rocky cliffs of this ridge descend to the river on the right bank from a height of about 75 sagens at an angle of 45°. On the left bank, which along the river presents a perfect plain, at a distance of about one verst, there appears a prolongation of the range in the form of hilly elevations, the whole range of hills being disposed in a semi circle across the stream. The Sheikh-Djeili ridge descends towards the river in two branches, and on the summit of the more southern branch, which does not reach right to the river, are to be seen the ruins of Fort Yamnuk-Kala. The other spur approaches close to the river and, between its steep slope and the water, there is only a narrow path. With the exception of this place, for the whole course of the river through the Khivan Khanate as well as in the delta, there are more or less permanent banks. The river bed, consisting of mud, in consequence of the absence of more solid parts remains unconsolidated, the result being a continual change of the river's course. The banks are, during the high water season, where they are not protected by dams, often inundated for a great distance, marshes and floods being formed.

The greatest changes of the banks take place in the month of September, when, the water levelbeing already very low, continues to still further fall. At this period of the year, along the whole excavated bank for about one or one and a half arshins from the top of the bank cracks are to be seen, which become larger little by little, the ground finally giving way and falling into the river, often in pieces some cubic sagens in size. On a calm September evening, the noise caused by the falling in of these masses of earth is constantly to be heard. In consequence of these properties of the ground, the very bed of the river, its channel as well as its sandbanks undergo considerable and frequent changes, often quite sudden. The flat left bank of the Amu-Darya, which forms the Khivan Oasis, is but very slightly raised above the water level so that, notwithstanding a comparatively very insignificant rising of the water, during the highwater season, ordinarily not exceeding 1.5 sagens, this bank is subjected to regular annual inundations, which would wash away the flourishing culture existing in that country, if special measures for preventing it were not taken. This is the reason why all inhabited places of the Khivan Oasis are protected from the inundations of the Amu-Darya by earth walls surrounding them, which are not more than one sagen high and about one and a half sagen thick.

The vast plain, occupied by the Khivan Oasis, consists of a muddy and sandy or alluvial soil, suggesting that that country was once nothing but the delta of the Amu-Darya, formed from the sediments, carried and deposited by the river. These sediments filled up that part of the Aral, which used once so be joined with the ancient Lake Sary-Kamysh, and to diminished the surface of that sea. The traces of old river beds, which scored this plain, prove that the course of the Amu-Darya's waters often changed there, as it does in the present delta, moving gradually from west to east, and it may be supposed that the time is not far off when the waters of the Amu-Darya, after filling up with their sediment all the low places of the present delta and finding in their eastward movement an ever slighter fall in the ground, will again choose the direction, where the inclination is greater, that is, that the Amu-Darya will again begin to flow westwards and will send its waters into the Sary-Kamysh basin.

In consequence of the hot climate of the steppe part of Central Asia and the lack of rain, agriculture is possible there only on condition of constant and abundant irrigation, the establishment of which requires a considerable quantity of labour. The irrigation works, annual repairing and cleaning of the aryks, are executed by all the owners of land under the inspection of specia lofficials called «aryk aksakals». Besides irrigation, the chief canals of the Khivan Khanate serve as convenient means of communication. Grain and other raw agricultural products are carried to the chief market places of the Khanate wholly by water, in local flat bottomed boats caledo «kayuks». The fall ot the water level in the chief canals is on an average 0.06 to 0.08 sagens per verst, and the depth during the highwater season about one sagen. In general, the chief canals are constructed tolerably well. The velocity of the current, in the season of ordinary water level, is not great enough to undermine he banks of the canals, although they are not strengthened in any way. Nevertheless, the current is strong enough to prevent the accumulation oi alluvium. The chief canals are cleaned very rarely and even then not throughont their whole length, but only in the widest places. The average velocity of the current during the high-water season is 3 to 4 sagens. The outlets of all the small canals from the chief canals are open, i. e., there are no arrangements for regulating the supply of water, in consequence of which the expenditure of water in the canals depends upon the water level in the Amu-Darya. Although, in general, the season of high water level in the Amu-Darya coincides with the season of the greatest expenditure of water upon irrigation, during the time of the highest water level in the Amu-Darya, that is, in the months of June and July, the supply of water to the canals is considerably greater than needed for that purpose. The greatest quantity of water for irrigation is required in the month of May. This circumstance proves the faultiness of the Khivan method of irrigation. In the highwater season, when the supply of water into the canals is greater than is necessary for irrigation, the water filling the canals undermines their banks and destroys bank protecting dams, causing every year no little work on the repair of bank dams and canals. There is equally no arrangement for directing a greater quantity of water into the canals in the month of May before the high water season, and it often happens that the supply of water to a canal in this season is not enough for the irrigation of the fields. Nearly all the canals issue from the Amu-Darya by several arms. This fact indicates that, at first, the canals branched off from the river by one arm, and only afterwards new ones were dug for the purpose of augmenting the supply of water. At present, in the highwater season, it is necessary to dam up one arm for the purpose of diminishing the flow of water.

The chief canals, which branch off from the Amu-Darya on the left side and carry water into the interior of the Khivan Khanate are the following:

1. The Polvan-Ata, taking its rise from the Amu-Darya 15 versts below the town of Pitniak. Its whole length is about 80 versts. At the end of it is situated the city of Khiva.

2. The Khazavat flows out of the Amu-Darya 14 versts below the exit of the Polvan-Ata. It is about 90 versts long.

3. The Shakh-Abat or Shavat flows out of the Amu-Darya opposite to the town of Khanki. It is the chief artery which supplies the Khivan Khanate with water, both as respects the quantity of water flowing through it, and its length, which is nearly 155 versts. The Shavat flows out of the Amu-Darya in two arms. That which has its exit higher up the river, is full only in the highwater season; it joins the lower arm 15 versts from the outlet of the latter from the Amu-Darya. Both arms of the Shavat, before they unite into one channel, traverse a low conntry, which being inundated during the highwater season, is always marshy. This is the reason of the deserted appearance of the strip of land next to the water. Only after some 5 versts, the country becomes comparatively higher, and cultivated patches of land are to be seen here and there.

The Shavat flows past the towns of New Urgench, Shakh-Abat-Tashauz and Ilialy. Below the last, it still flows some 20 versts, and then becomes shallow, distributing its water to the adjacent fields, and vanishes altogether.

4. The Yarmysh flows out of the Amu-Darya some 15 versts below the exit of the Shakh-Abat; its length is 79 versts.

5. The Klych-Niaz-Bai flows out of the Amu 8 versts below the preceding canal, and is 102 versts long.

6. The Mangyt-Arna flows out of the Amu 60 versts below the Klych-Niaz-Bai and has a length of 82 versts.

7. The Lauzan branches off from the Amu-Darya considerably lower down, beyond the ridge of Sheikh-Djeili. The general direction of the Amu-Darya's course through the Khivan Khanate is northwest but, from the Kosh-Tiube elevation on the right bank of the river, it changes to southwest. Afterwards, the river takes a sharp bend, and again flows northwest to where its delta begins. Just at this sharp turn of the Amu-Darya, in its deepest point, the Lauzan branches off to the left and flows nearly normally relative to the general course of the Amu-Darya. Thus, its situation presents a very favourable condition for the waters of the Amu-Darya to flow by it, making out of the Lauzan so to say a prolongation of the chief river bed, with the same direction and an identical fall. At present, the Lauzan canal branches off from the Amu-Darya in three arms, of which the highest up the river is called Esaul-Bashi, the middle one Divan-Begi and the lowest little Lauzan, previously Kaun-Uziak. Five and a half versts from their exit, all the three named arms unite. In the fourth arm, the Great Lauzan, a dam has been constructed called Bent. The Khivan Khans, from motives of policy, not wishing to let too much water out of the Amu-Darya to the Turkomans dwelling along the Lauzan and lower dam along the banks of the old river bed, the Kunya-Darya, dammed up the bed of the Lauzan at its exit out of the Amu-Darya with a solid structure, only a very insignificant quantity of water being let into the canal, for the irrigation of the nearest fields, for which purpose two little arms, called Divan-Begi and Esaul-Bashi, were dug to avoid the obstacle erected. In order to observe the condition of the said dam, as well as to control the quantity of water permitted to be let into the Lauzan, Fort Bent was built there, the

ruins of which are to be seen even to day. In consequence of this artificial lowering of the water level in the Lauzan, its general character, in the parts better preserved, is that of a double profile, the higher part having a great width and comparatively insignificant depth, and the lower one representing so to say a second and deeper bed, by which water flows with a narrow stream between the second ary banks lying in the old wide bed. From the point of junction of the Great and Little Lauzan, the water flows for a distance of about 5 versts and then runs into the canal Saubet-Yargan, which represents as it were a continuation of the Lauzan and receives the greatest part of its waters. Below the Saubet-Yargan, on both banks of the Lauzan, traces of a dam are to be seen. Beginning with the fifth verst, the left bank of the Saubet-Yargan is accompanied with floods, which receive their water from the overflowings of the canals Mangyt-Arna and Klych-Niaz-Bai. They are overgrown with cane. Approaching the old river bed, the Kunya-Darya, into which the Saubet-Yargan lets its waters, the country rises somewhat. Before the spot where the Saubet-Yargan falls into the Kunya-Darya, the canal Murtaza-Yab (Yakub-Yargan) separates to the right, and the Katta-Kar to the left. Both of them fall into the Kunya-Darya.

After the branching off of the Saubet-Yargan, the Lauzan becomes at once narrower and shallower. Throughout its whole length, as far as the outlet of the Nurak canal, the banks are visible and rise above the water level, but in many places they are low, overgrown with bushes, cane, tamarisk thorny bushes and djida. Below the Nurak, the Lauzan is again dammed up. The Nurak is at present also a kind of prolongation of the Lauzan. Some versts before the outlet of the Nurak from the Lauzan, there begin uninterrupted floods, known under the name of Kazetan-Kul or Kara-Kul, overgrown with cane. They stand in communication with those of the Saubet-Yargan. These lakes, or properly speaking marshes, extend for about three versts. From them flow two channels which, before falling into the Kunya-Darya, join into one bed, which falls into this old channel.

Below the dam at the outlet of the Nurak, the Lauzan has no stream for about a verst. Farther on, it obtains water from the Yakub-Yargan canal, which flows out of the Suelli canal, which, in its turn, quits the Amu-Darya below the Lauzan at about 20 versts from the town of Khodjeili. To the dam at the outlet of the Khan-Yab canal, irrigating the district of Kunya-Urgench, the Lauzan flows through a ravine between double banks, those of the now existing canal and the former Lauzan, whose bed was considerably wider and part of which is silted up with sand.

Below the dam at the exit of the canal Khan-Yab, at some distance from it, another dam was constructed of fascines; the part of the bed between the two dams was gradually filled up by the sediment carried there by water, flowing by an insignificant encircling aryk.

8. The canal Shamrat flows out of the Kunya-Darya before the Shamrat dam, and has a southwesterly direction. As was mentioned above, the waters of the Lauzan, after traversing the floods of lake Kara-Kul by the canal Katta-Kar, fall into the old bed of the Kunya-Darya, which also receives the supperfluous water of other aryks. For this distance, the bed of the Kunya-Darya is completely full of water and has the appearance of a river. In proportion to its length and to the quantity of water it carries, the Shamrat irrigates only a very insignificant surface, properly speaking a mere strip of land on its right bank. The land, on the left bank of it, is irrigated by the water of the Divan-Begi-Yargan, which detaches itself from the Kunya-Darya at the dam of Kizil-Takyr, so that only now and then small patches of land near the left bank of the Shamrat profit by the waters of that canal. In consequence of a comparatively great elevation of the lands situated in the immediate vicinity of the canal, fields begin to be cultivated only where they are situated at a considerable distance from the canal, where their level falls, thus making the irrigation easier. Farther on, inhabited places disappear, and the bed of the canal traverses a sandy desert. Then, the Shamrat parts into two branches, each of which receives an equal or nearly equal quantity of water, flowing by the canal to this point.

From this fork, cultivated fields again begin to appear. After passing the old bed of the Daudan, the Shamrat canal begins to diminish gradually and the current of water in it stops.

The Kunya-Darya.

In the Khivan Khanate special attention is attracted by the old beds of the Amu-Darya, i. e. the beds, by which this river used to flow in ancient times. The Kunya-Darya is the most remarkable of these old beds. It leaves the Amu-Darya near New Urgench, between the canals Yarmysh and Klych-Niaz-Bai. The narrow stream of water, trickling through the bed of the Kunya-Darya, is called by the natives «Daryalik», which means «little river», to distinguish it from the bed, in which this stream lies, and which is called Kunya-Darya «old river».

The water of the Amu-Darya reaches the Daryalik in a very limited quantity, viz. 3 to 3.5 cubic sagens during the highwater season. The chief destination of this water is to irrigate the localities, situated along the canal Kizil-Takyr, which leaves the Kunya-Darya above the Kizil-Takyr dam. After quitting the Amu-Darya, the water flows by the bed of the Kunya-Darya in a very irregular stream, making continuous turns and so-called knees. Farther on, the bed is not so obstructed with mud and the water overflows the whole valley, so that the Kunya-Darya has the appearance of a lake. The width of the bed is 100 to 250 sagens and occasionally increases to 2 versts. The Kunya-Darya retains this character up to the dam of Kizil-Takyr or Divan-Begi. Below this, the bed of the Kunya-Darya is waterless, and has the appearance of a dried up river bed with a regular transverse section, about 120 sagens broad, although even there it is occasionally stopped up with sandy deposits. Passing by villages situated near to it, water reappears now and then in the Kunya-Darya bed but, after receiving the waters of the Lauzan, it again has the aspect of a river up to the Shamrat dam. Beyond this, water appears only in summer, at that time of the year the water, which remains in the canals after the irrigation, passing into the Kunya-Dárya. After the dam of Medemi, the bed of the Kunya-Darya again has the aspect of a dried up river, which it retains right up to where it falls into the basin of Sary-Kamysh. From the dam of Medemi, the actual bed itself has a very regular form. Sandy deposits, formed after the year 1878, are to be seen but very rarely. Covering the banks, and the slope and the bottom of the bed, they diminish its transverse section considerably.

The whole course of the Kunya-Darya bed from its quitting the Amu-Darya to the level of Lake Sary-Kamysh has a length along the channel of 436 versts. This whole distance may be divided, as regards its character, into three parts. The first part abundant in water, from the Amu-Darya by the channel up to the Kizil-Takyr dam is 114 versts long. The second part, from the Kizil-Takyr dam to the town of Kunya-Urgench, is 99 versts in extent, and has a stream of water only for some distance before the dams of Shamrat and Medemi. The third and last part, from Kunya-Urgench to the Sary-Kamysh, 222,5 versts in length, presents a dried up river bed, which on former maps was called Urun-Darya. It has a tolerably monotonous character with the exception of the part in the immediate neighbourhood of the lakes of Sary-Kamysh, which has undergone considerable changes during the last period of its existence after the fall of the level of the Sary-Kamysh lake once existing there.

The last part of the Kunya-Darya's bed is, on the whole, tolerably deep and comparatively clean, although even here it is in some places silted up with sandy deposits, which diminish its transverse section. In proportion as it flows westwards, the width of the Kunya-Darya diminishes, and the depth of the bed increases considerably. There is no more sand in the bed; its bottom is clayey and tolerably solid; its banks become almost quite perpendicular. This last characteristic of the Kunya-Darya's bed begins to prevail in the part contained between the hills of Butenau and 30 versts before the present lakes of Sary-Kamysh by the channel. It there attains a width of 60 to 70 sagens, with the banks rising above the bottom of the bed for more than 20 sagens.

As to the Kunya-Darya's bed for the last 40 versts of its course, the banks here become rapidly lower. After a distance of about 20 versts, they have a height of about 2 sagens and farther on they disappear completely.

There are 8 dams in the bed of the Kunya-Darya from its exit out of the Amu-Darya to the Sary-Kamysh basin, situated along the longitudinal section of the bed as follows:

1. At the 101-st verst — Kizil-Takyr.
2. » » 139-th » » Shamrat.
3. » » 156-th » » Medemi (Mahomet-Vali).
4. » » 158-th » » Tadjik-Bent.
5. » » 196-th » » Ushak-Bent.
6. » » 205-th » » Kizil-Chege.
7. » » 216-th » » Salak-Bent.
8. » » 242-nd » » Egen-Klych.

As to the fall of the bottom, throughout the whole length along the channel, from where it quits the Amu-Darya to the level of the water in the present lakes of Sary-Kamysh, it may be divided as follows. The elevation of the bottom of the Kunya-Darya bed at its exit from the Amu-Darya, above the water level in Lake Sary-Kamysh was 62.30 sagens, this height being divided as follows: 1) the part of the bed from the Amu-Darya to the outlet of the canal Kizil-Takyr, at the dam bearing the same name, has a fall of 8.62 sagens to 114.38 versts of the channel, which makes 0.075 sagen per verst; 2) from this place to the inlet of the Katta-Kar canal, 2.76 sagens for a distance of 39.16 versts, or 0.07 sagen per verst; 3) from the inlet of the Katta-Kar canal to the issueof the Shamrat, the bottom of the bed has an inverse fall of 0.09 sagen to a distance of 12.58 versts, which is equivalent to 0.007 sagen per verst; 4) from the outlet of the Shamrat canal to the inlet of the Lauzan, there is a fall of 2.12 sagens to a distance of 32.71 versts, or 0.065 sagen per verst; 5) from the inlet of the Lauzan to the point opposite to the elevation Beten-Dag there are 6.39 sagens to a distance of 173.35 versts, or 0.037 sagen per verst; 6) to the site of Dekcha, there are 0,98 sagens for a distance of 13.48 versts, or 0,073 sagen per verst; 7) from the site of Dekcha to the wells of Sary-Kamysh there are 30.65 sagens to a distance of 40.80 versts, or 0.75 sagen per verst; and 8) from the wells of Sary-Kamysh to the level of the lake of the same name, 10.87 sagens to a distance of 9.59 versts, or 1.133 sagen per verst; for the whole distance the average fall is 0.1436 sagen per verst, and the incline 0.0002858.

To conclude the general description of the Kunya-Darya bed, it may be stated that, for a distance of 436 versts from where it leaves the Amu-Darya, it has no considerable arms. It neither receives nor sends out any channels but artificial ones. It is only in its lower part that an arm called Kichkine-Darya or «little river» branches off from it, so that the Kunya-Darya falls into the Sary-Kamysh basin through two arms. As to the nature of the ground, in the localities traversed by the Kunya-Darya from the Amu-Darya to Kunya-Urgench, it presents the general type of a muddy clayey soil, in general characteristic of the whole area of the Khivan oasis.

The same muddy clayey soil extends farther for about 80 versts beyond the frontier of the Khivan Khanate in a narrow strip of fertile land

But this whole region is now deserted and has the aspect of a perfect wilderness, bearing traces of former civilization, of an extensive agriculture and a tolerably dense population, as proved by the ruins of buildings, whole towns and forts.

The Daudan.

After the Kunya-Darya, the Daudan is the second of the Amu-Darya ancient beds, traversing the Khivan Khanate, which deserves our attention. There are now no exact indications as to the former outlet of the Daudan from the Amu-Darya. The natives call a ravine, which begins between the Shakh-Abat and Khazavat canal seven versts below New-Urgench, the beginning of the Daudan. A little aryk flows along this ravine, which leaves the Shakh-Abat canal on the left about 26 versts from the Amu-Darya. As the Daudan at present receives the greater part of its waters from the Shakh-Abat canal by means of this aryk, and as in general an increase of the supply of water to the Daudan can be effected best of all by this canal, the first 26 versts of this canal may be said to represent the beginning of the Daudan. For the first 18 versts from the Shakh-Abat canal, the bed of the Daudan offers a valley whose transverse inclination is imperceptible to the eye, along which flows an extremely crooked aryk. The breadth of this ravine is about one verst. The nature of the soil is muddy and clayey. Afterwards the bed of the Daudan enlarges, reaching a width of 2.5 versts, and gradually passes into marshes and lakes of most varied outlines, overgrown with dense cane and joined with each other by narrow streams. The water forming these floods flows into the Daudan out of the Shakh-Abat and Hazavat canals, this bed serving at present as an outlet for the superfluous water remaining from the irrigation of the fields. Therefore, the greatest overflowing of the Daudan does not stand in relation to the season of highwater in the Amu-Darya, but to the time when the irrigation of the fields ceases.

Near to the village of Igdyr-Kala, the bed of the Daudan divides into two arms, which join again into one channel 60 versts farther on at the Pitnek-Bent dam. The whole quantity of water flows by the left arm, which in no way resembles the bed of a formerly existing river; the outlines of the ancient river bed are perfectly effaced, the bed of the Daudan now presenting a series of lakes and floods of the most various forms, joined by narrow and insignificant streams. The right arm of the Daudan has from its outlet, at least at the beginning, more or less the aspect of a river bed, although the action of winds and the water have considerably altered the outlines of its banks and bottom. Farther on, the character of the bed is somewhat changed; it loses its regularity and presents in summer a succession of floods and sheets of water. The outlines of the banks then become again somewhat more definite, and in general the right arm acquires the appearance of a river bed although an extremely irregular one as far as width is concerned. Throughout this whole distance, the country from the Daudan to the Shakh-Abat canal is very well cultivated and presents, as it were, an unbroken garden.

Farther on, along the right bank, corn fields and gardens alternate with occasional saltmarshes or bushes of tamarisk, while the left bank is sandy and devoid alike of vegetation and habitations. After the junction of both arms below the Pitnek-Bent dam, the bed of the Daudan has at first the character of a lake, but farther on it again narrows. At the outlet of the former Chermen-Yab canal, the section of the Daudan containing water ends.

Throughout the whole length of the upper part of the Daudan, the bottom and waterlevel are lower, than they are in the adjacent Shakh-Abat and Khazavat canal, the difference reaching one sagen.

Below the outlet of the Chermen-Yab canal, the bed of the Daudan is completely silted up with sand, and level with the country it traverses, so that all traces of it are lost. Here is a succession of winding valleys or depressions, which extend in a northwestern direction nearly up to the Shamrat canal. These valleys do not form a continuous river bed, but stretch in a series of isolated dells and closed ravines, obstructed by great shifting sand dunes devoid of all vegetation and hence without any cohesion. On approaching towards the Shamrat some five or six versts, these masses of sand begin to decrease both in size and as regards the surface covered by them, and a more regular and continuous river bed with true banks composed of a more solid clayey soil begins to show more distinctly. In the immediate neighbourbood of the Shamrat, only some 150 to 200 sagens before reaching it, the bed of the Daudan becomes lost again and completely levelled with the surrounding country. The same phenomenon is to be observed on the other side of the canal, where no traces of the bed are to be found, it beginning again only about 300 sagens from the Shamrat. After a course of about 4 versts, a tolerably regular and clearly defined river bed reappears with all the symptoms of an ancient river, sending off a few lateral, now dried up aryks. About 6 versts before the fort Mangyr-Kala, the bed of the Daudan parts into several arms, which at one time unite with each other and at another lose themselves in the adjacent plains. Throughout the above stated distance, a sandy soil predominates, while the vegetation consists of dense brushwood overgrowing the banks and bottom of the actual bed.

The part of the Daudan, from the fort Mangyr-Kala to the end, presents in general the greatest regularity. The chief feature of this section of the Daudan, distinguishing it from those following, is the circumstance that a small bed of about 12 sagens in breadth and about one sagen in depth runs along its chief bed. This secondary bed, as it may be called, is nearly everywhere visible. According to the statements of the natives, it was formed in the Daudan when water flowed through it on the last occasion, namely, when water flowed into the Daudan from the Shamrat canal. From Mangyr-Kala to the wells of Kudju-Kuyu, there are four dams in the bed of the Daudan at distances of 15, 20, 22 and 38 versts from Mangyr-Kala. The last of them bears the name of Kum-Bugut. At the Kudju-Kuyu wells the Daudan parts into two arms, of which the right for the first two and a half versts has the features of an old river bed; by it probably the water was emptied into Lake Sary-Kamysh. The left arm was at first only an aryk, dug for the purpose of carrying water from the Daudan to irrigate the country situated to the northwest of the Sary-Kamysh. Only subsequently, this aryk was washed away by the water let into the Daudan. This arm leaves the main bed nearly at a right angle, and one and a half verst lower down is lost in a depression, which is the bottom of an ancient lake. This dried up lake also extends in a southern direction; its length is about 10 versts, and breadth from 3 to 4 versts, while the depth for the most part does not exceed half a sagen. The bottom of this depression presents a completely level surface, covered with dense tamarisk bushes. Farther on, beyond this dried up lake, the river bed begins again, and for the first two versts takes a southwestern direction, and then turns to the south to the Baster-Molla wells. For this distance the bed is more or less regular, but to a great extent silted up with sand. Below the Baster-Molla wells, the bed divides into small arms, which afterwards join again into one channel, in its turn lost in Lake Tiuniu-Klu.

This dried up Lake Tiuniu-Klu has at present a round form; its surface is about 15 to 16 square versts, its depth about 6 sagens. The bottom is an even plain and its banks are surrounded by sand hills. In the northwestern part of Lake Tiuniu-Klu, a river bed leaves it whose outlet is completely filled up with sand, but which somewhat farther on becomes visible. It runs west and is lost in the basin of the Sary-Kamysh. It may be supposed that this bed was formed, when the water of the Sary-Kamysh basin began to fall and to retreat westwards, so that Lake Tiuniu-Klu became severed from the common basin; hence it is that the depth of the bed increases to the west. For the whole length of the Daudan, traces of a past civilization are visible. Such are the ruins of forts and «saklias» or houses, old aryks silted up with sand, etc. The country now and then is covered with thinly growing and low saksauls and small sand hills, situated for the most part on the banks of abandoned aryks. So that, even at present, if water should be let into the Daudan or the Kunya-Darya, this country would prove to be fit for agriculture.

The Sary-Kamysh basin.

Those of the old beds of the Amu-Darya, which leave it on the left bank and take a westerly direction, once discharged their waters into Lake Sary-Kamysh, which now presents a dried up depression. The bottom of the ancient lake has at present the following appearance. The eastern part of it, for a distance of 20 to 25 versts westwards from the Tarym-Kaya hills, northwards to the end of the Daudan and southwards to the Kanga-Kyr, is an even nearly horizontal surface, the height of which above the level of the Sea of Aral varies from zero to one and a half sagen. This section is overgrown with low saksaul, and is covered now and again in its north-eastern part with small sand hills, situated mostly along the ravines existing there, which have a southwestern direction. The deepest place in the basin is in the northern part of it. At present there are two salt lakes, known under the name of the Sary-Kamysh Lakes. They were previously joined by a narrow channel about 50 sagens wide, which dried up and turned into a patch of semiliquid salt mud. Lengthways these lakes extend from north to south. Their surface occupies about 130 square versts. The water level of the lakes is lower than that of the Aral by 42.11 sagens and, as in its turn, the level of the Caspian is lower than that of the Aral by 36.39 sagens, the level of these lakes proves to be 5.72 below that of the Caspian. The greatest depth in the lakes is 2.5 to 3 sagens, their banks are slanting, especially the case with the eastern and south-eastern. The bottom consists of salt mud covered with a crust of solid salt. A strip of the banks near the water line presents a very sticky, quite impassable marsh. At a distance of one verst from the surface of the Lakes of Sary-Kamysh, the locality rises 6 to 10 sagens, and further remains in general horizontal, so that the bottom of the ancient lake represents an immense plain extending southward for about 30 versts, eastward about 8 versts, westward about 15 versts from the level of the now existing lakes. On this plain there are to be seen here and there elevations, both rocky and sandy, which were formed after the drying up of the lake. Deposits of precipitated salt are only to be met with in the vicinity of the now existing lakes, their thickness being insignificant. This points to the fact that the water of the ancient Lake Sary-Kamysh contained a trifling proportion of salt, so that, were the lakes of Sary-Kamysh filled with water, it may be supposed that the latter would not be saline. There exists yet another proof that the water of the ancient lake was fresh. On the western slope of the basin, beginning from the road which leads from the Sary-Kamysh wells to those of Uzun-Kuyu, on the Ust-Urt, southwards, numerous traces are met with of ancient irrigation aryks and of holes used for placing the chigirs or water lifting wheels, in a locality situated 16 to 18 sagens below the level of the Sea of Aral. The guides related, and their statement is deserving of credit, that agriculture used to flourish there, and that the water of the lake was used for the irrigation of the fields on its shores. Consequently, even when its water level had fallen from 18 to 20 sagens, the water of the lake was sufficiently fresh to be fit for the irrigation of the fields. The supply of spring water to the basin is at present very trifling. Only in the northern and northwestern parts, wells and springs are to be met with, fed by water from the Ust-Urt. But their number is only insignificant, the supply of water being far from sufficient to cover the annual loss of water by evaporation from the surface of the present lakes. The level of the latter 6 falling every year, and the surface of the water diminishing constantly, so that very soon these lakes will be transformed into a salt marsh.

On the eastern slope, there are no springs at all. Some twenty years ago, according to the account of the guides, in the wells of Aman-Shiikh, situated at the bottom of the now dried up Lake Tiuniu-Klu, water was always to be found at a depth of 5 to 6 sagens. During the operations of the Expedition, one of these wells was deepened for 20 sagens, or 25 sagens below the level of the Aral, but no water was found. On the whole, a falling of the water level is to be observed in the wells situated eastwards from the Sary-Kamysh basin, for example in those of Kudju-Kuju, Baster-Molla and Giaur-Kala. During the last twenty years, as the guides assured us, the fall of the water level in them has reached 5 sagens.

Vegetation in the basin is scanty, and in general to be met with only around the lakes of Sary-Kamysh and on their eastern edge. In these places it consists of saksaul and tamarisk. The bottom and banks of the dried up Lake Tiuniu-Klu are covered with an extraordinarily dense thicket of tamarisk, which also proves that not so very long ago fresh water must have existed there Over the whole remaining space occupied by the Sary-Kamysh basin, the vegetation consists of scanty saksaul hardly rising above the surface of the ground. Among the lakes and sheets of water which form the southern part of the Sary-Kamysh basin, especial attention is due to the system of saltmarshes and dried up lakes, forming the western bay, which extends southwards from the Charyshly wells in the direction of the site of Kugunek-Kyr, the beginning of the Uzboy. This system represents, as it were, a direct continuation of the Sary-Kamysh basin, out of which the Uzboy flows. The saltmarshes, situated to the south of Zengi-Baba in the direction of the Shiikh wells, were partly separated from the chief basin already at the epoch when the Lake Sary-Kamysh existed, and will in no case be flooded in the future, because the level of the projected future lake will be lower than was that of the ancient lake.

From the Charyshly wells to the site of Siradja, the western gulf of the ancient Sary-Kamysh basin presents a flat valley, bounded on the west by gently sloping spurs of the Ust-Urt, on the east by high ridges of sand hills, situated on a clayey sandy soil. The height of these hills is such that their summits are on a level with the present Sea of Aral.

The width of this valley, taking into consideration the former level of the lake, was from 5 to 15 versts. Its bottom is covered with low sand hills. Along the middle of it runs another valley, which has the appearance of an irregular river bed. The width of this bed at the Charyshly wells is one and a half verst. Moving southwards from the Charyshly wells, the breadth of this bed diminishes and, about 30 versts below Charyshly, it does not exceed 120 sagens, which remains unaltered for about 10 versts. For this distance, it has the features of a perfectly regular river bed. It then again widens, attaining a breadth of 400 sagens, and is lost in the common valley of Siradja. The depth of this bed throughout its whole length is 3 to 4 sagens. From Siradja, the general character of the valley remains the same but, in the place of the river bed in the middle, there is a succession of saltmarshes and sheets of water of the most varied form, joined together by narrow paltry streams. The breadth of the valley is the same as in its preceding section, 8 to 12 versts. The width of its deeper parts is 1 to 4 versts its surface being there covered with less numerous sand hills. The soil is of a muddy and sandy nature, with an addition of particles of chalk. Throughout nearly the whole length of the valley, a great quantity of shells are scattered upon its surface, among which fresh-water specimens occur. Many traces of decayed cane and even deposits of peat were met with. At a depth of 0,2 sagens, the latter was found on the site of Siradja, and 10 versts above the beginning of the Uzboy.

It may be supposed, from the existence of spots, where fresh-water shells are found, as well as from the situation of the bottom of the first part of the Uzboy, which once served as a channel for the passage of the waters of Lake Sary-Kamysh into the Caspian, that the level of water in the ancient Lake Sary-Kamysh, at the epoch of the existence of communication between the Amu-Darya and the Caspian, was one and a half to two sagens higher than the present level of the Sea of Aral. The limits of the ancient Lake Sary-Kamysh were probably as follows. To the east, it extended as far as the Nurum-Kry and Tarym-Kaya hills; on the south-east it reached the Ishek-Ankren-Kyr hills, washing the western slope of Mount Zengi-Baba; on the north and west, the ancient Lake Sary-Kamysh was bounded by the plateau of Ust-Urt. As to the southern limit, it is not easy to state it exactly at present; the hypothesis may be hazarded that it was formed by the north-western spurs of the Ishek-Ankren-Kyr and the sandy hills of Gok-Navat, and Ak-Navat, or that the lake extended nearly as far as the parallel passing through the Charyshly wells. Farther to the south, a succession of lakes and sheets of water presented a continuation of the ancient lake. They were joined with each other and with the basin of Lake Sary-Kamysh.

These systems of lakes formed as it were three gulfs, and are situated as follows. The most western extends from the Charyshly wells south in the direction of the site of Kugunek, where water flowed out of it into the Uzboy, and so on into the Caspian Sea. The next system of dried up lakes is insignificant in extent. It is bounded on the west by Gok-Navat, on the east by Ak-Navat and the western spurs of the Ishek-Ankren-Kyr heights. These lakes, united with Lake Sary-Kamysh by narrow shallow gulleys, extended southwards to about the Nefes-Kuyu wells.

This is confirmed by an examination of the locality and levels taken nearly 15 versts north of the site of Kugunek in an eastern direction, approximately 10 versts east of the meridian of the Orta-Kuyu wells. This levelling showed that the whole locality, situated to the east of the saltmarshes of Siradja, the western gulf of the ancient Lake Sary-Kamysh, is covered with sand hills, the height of which is about 15 sagens above the level of the Sea of Aral, and that the lowest points of the bottoms of the small hollows, situated between the sand hills, are all above the level of that Sea. These sand hills are bounded to the east by a rocky elevation, which extends from north to south and is probably a continuation of the Ishek-Ankren-Kyr hills. Finally, the third and most eastern system of lakes, forming the southern part of the ancient Lake Sary-Kamysh, consists of saltmarshes, situated on the southern side of the Kanga-Kyr hills, and extending in a south-easterly direction to the western slope of Zengi-Baba. Hence they extend farther in a south-easterly direction and reach the Shiikh wells. The saltmarshes, situated westwards from the Shiikh-wells in the direction of those of Bala-Ishem on the banks of the Uzboy, known also under the name of Unguz, were in communication with the ancient Sary-Kamysh basin only by the above named saltmarshes. According to the levelling, and detailed examination of the country between the Bala-Ishem wells and Cape Kizil-Murun, these saltmarshes have no communication with those stretching from the site of Siradja to the Uzboy, nor with the Uzboy itself.

The valley of the chief Unguz, which extends from west to east, is on the north joined by a series of similar but smaller valleys forming lateral Unguzes. Near the Shiikh wells, the country is covered with high shifting sand hills, which are interrupted only at the wells themselves, situated on a little plateau, surrounded on all sides by sand. Over the whole area above described neither shells nor river formations are met with. The soil is nearly everywhere sandy, marl occurs less frequently. The latter forms the bottom of the valleys, and has the appearance of red clay washed by water and interspersed with crystalline gypsum.

Gypsum sandstones and limestones constitute here occasionally independent formations found even in transverse ridges barring the valleys. Under the upper layers of the saltmarshes, locally called «sor» or «shor», at a depth of about half an arshin, pure coarse grained red ferruginous sand was found. Thus, the levelling executed for a distance of more than 53 versts throughout the series of hollows of the Unguz, as well as the topographical character of the whole locality explored between the Uzboy and the Shiikh wells, the absence of muddy river deposits and shells in the valleys, and their very form, all these facts prove that the Unguz cannot be regarded as a bed of the Amu-Darya and that consequently no flow of water by it into the Uzboy was possible, as was formerly supposed. Between the Bala-Ishem wells, or Orta-Kuyu, and Islam-

Kuyu, as also to the east from the Shiikh wells for a distance of 20 to 30 versts, no traces of a river bed or even of water having flowed there are to be found. The bottom of the Unguzes, between the Islam-Kuyu and Ata-Kyu wells, is lower than that of the Uzboy at the Bala-Ishem wells by 1 to 2 sagens, so that the hypothesis of the Amu-Darya having flowed by the Unguz is absurd.

The space between the systems of saltmarshes and hollows described above is for the most part covered with enormous sand hills, the summits of which rise about 15 sagens above the level of the Aral. Especially high are the hills situated over an area of 25 to 30 versts to the east of the western system of saltmarshes. These sand hills are for the most part disposed in ridges extending from north to south, with an inclination to the northeast of 10^0 to 15^0.

The Amu-Darya fell formerly into Lake Sary-Kamysh probably on the south-eastern side, by the bed of the so called Chardjui. Water formerly flowed by this channel into a system of lakes between the Shiikh wells and those of Laily, then gradually turned eastwards along the southwestern slope of Zengi-Baba to the depression of Ide-Khauz, which it filled up with its deposits. In consequence of this accumulation, the stream was gradually displaced to the north-east and undermined the hills of Zengi-Baba and Kanga-Kyr and the mound of Koy-Kyrlgan, which in all probability once formed an unbroken elevation. The southern part of the Zengi-Baba and Kanga-Kyr testify that great depredations were commited by water.

These elevations represent, on their southern side, perfectly perpendicular rocks 25 sagens in height, while Kanga-Kyr and Zengi-Baba sink gradually, the former to the north and west, the latter to the north and east, and reach finally the level of the surrounding country. All these elevations consist of lime formations, which are easily destroyed by the action of water. The undermining on one side, and the deposits on the other, by the waters of the Amu-Darya, explain also the origin and present condition of the so called Kanga-Darya or Tonu-Darya, which runs along the foot of the southern slope of the Kanga-Kyr hills, and consists of four separate channels, differing from each other in width, depth, and elevation of the bottom, and uniting the three basins.

Together with the change of the direction of the Amu-Darya's course in its upper reaches, its outlet into the Sary-Kamysh changed also. It was displaced to the north, and the Amu-Darya began to discharge its waters at first into the Daudan and then into the Kunya-Darya. The alluvial soil, which fills up the whole eastern edge of the Sary-Kamysh up to the northern spurs of the Kanga-Kyr Heights, testifies to the length of time during which the Amu-Darya ran in this direction.

The Uzboy.

In ancient times, the Uzboy served as the only outlet of the waters of the former Lake Sary-Kamysh. The rise of the Uzboy, as in general the source of a stream flowing out of a water basin, is not easy to define exactly. In such cases, the lacustrine character of the water mass changes gradually into that of a river. Below the mountain Gech-Geldy, in the saltmarshes of the western gulf of the ancient Lake Sary-Kamysh, the bed of the Uzboy begins to appear, at first indistinctly and imperceptibly traced in the form of a valley of an irregular shape and with an indefinite inclination; but, as the mountain Kugunek-Kyr, which the bed of the Uzboy surrounds on three sides, is approached, it obtains the normal features of a river bed, and has a regular inclination in the direction of the Caspian, so that this place may be considered to be the beginning of the Uzboy.

The total length of the Uzboy, from Kugunek to its issue into the depression of Bala-Ishem on the Balkhan Gulf of the Caspian Sea, is about 510 versts. For this whole distance, it has a very sharply defined river bed. As to the elements of the bed, its width, depth, and fall, they are extremely varied in different parts of the Uzboy. This variety depended chiefly on the state of the locality, formation of the ground and the inclination, through which the water was forced to cut its way. To the secondary causes of the irregularities of the Uzboy in different parts of it, may be reckoned the fact that, in proportion to the drying up of the Sary-Kamysh basin, the supply of water to the Uzboy constantly diminished, and its bed dried up gradually and was partly silted up by deposits, beginning at its outlet. The fall, on the whole 510 versts of its course, is about 35 sagens and equivalent on an average to about 0.07 sagen per verst; that is to say, the average fall of the Uzboy is equal to that of the Amu-Darya between Petro-Alexandrovsk and Nukus. But, as may be seen from what follows, this fall is very unequally distributed between the different parts of the Uzboy. As regards breadth, depth and particularly the distribution of the fall, the Uzboy may be divided into the following four sections:

1. From the mountain of Kugunek to the end of the Akh-Yaily cataract, a distance of 95 versts.

2. From the Akh-Yaily cataract to the Tagalek wells, a distance of 107 versts.

3. From the Tagalek wells to the spring of Dekcha, a distance of 53 versts.

4. From Dekcha to the issue of the Uzboy in the salt marshes of the depression of Bala-Ishem, 20 versts below the Tanderly wells, a distance of 193 versts.

From Kugunek-Kyr, the bed of the Uzboy has a southern direction, and retains it to the Kurtysh wells. Two versts before them, it makes a tolerably abrupt turn to the right, and for a distance of 4 versts, to the cataract Kurtysh, has a western direction. Below the cataract, the bed makes an insignificant bend and trends northwest to a cemetery, situated 5 versts below the Akh-Yaily wells. At this cemetery, the bed makes a sudden turn to the left and, to the end of the Akh-Yaily cataract, maintains a southwesterly direction. With the exception of these turns, and the bend which the Uzboy makes 10 versts after its exit, running round the western spurs of the Kugunek-Kyr, its bed is in general a straight one, especially in the section from the wells of Kurtysh to the cemetery of Akh-Yaily. The distinct characteristic of this part of the Uzboy is a considerable width; a great fall and a firm bottom. As the Ust-Urt in general consists of rocks which easily succumb to the destructive action of water, the Uzboy, as it flowed, succeeded in cutting itself a tolerably regular bed, both as to width and depth, with the exception of places where the water happened to meet with more solid rocks. Here it had not time to destroy them, and these rocks, crossing the river bed in little ridges, form a succession of cataracts, such as Kurtysh and Akh-Yaily.

Six versts before the Kurtysh wells, the bed of the Uzboy narrows down to 60 sagens while the banks rise to 5 or 6 sagens, consist of a clayey soil, and become steep and broken. The bottom of the bed is covered with a layer of white salt. The bed of the Uzboy retains this character for a distance of 8 versts, to the cataract of Kurtysh. Three versts below the Kurtysh wells, the Uzboy is crossed by a range of solid limestone rocks of the Ust-Urt, which form a cataract, like a dam constructed across the bed, three sagens high and three versts in width at its base, and one at its summit..

This cataract descends by three steps about 50 sagens long to a salt lake, which begins at the foot of the third. These terraces are perfectly perpendicular, and the height of the upper two is 0.5 sagen each, and that of the lowest 2 sagens. The depth of the salt lake near the cataract is about 7 sagens, a fact which indicates its formation by the bed being washed away by the pressure of water flowing in it. Below the cataract of Kurtysh, the bed of the Uzboy widens to 200 sagens, with a depth of 1 to $2^1/_2$ sagens, and remains such to the Akh-Yaily cataracts. These cataracts are formed by an insignificant elevation of the bottom above the surrounding surface. The bottom however consists of rocks, while the banks are of a clayey soil. These cataracts do not descend, like those above described, by steps but gradually for a more considerable horizontal distance, and form a kind of waste weir.

The following part of the Uzboy differs from the preceding. Its chief distinction is a slight fall, the uniform width of its bed for the whole distance of its course, and a considerable depth. For this distance, the Uzboy is bound in mostly between clay banks and a bottom of the same formation; only now and then, its bed is crossed by separate ridges of spurs of the Ust-Urt, which are almost washed out by the water, so that those places present no obstacle at all to a future stream of water by the Uzboy, with the exception of the rocky ridge running across it near the Lower Igdy wells, which is only to a small degree affected by the action of the water pressure. The general direction of the Uzboy, from the Ak-Yaily cataracts to the Igdy wells, a distance of 42 versts, is a southwesterly one. Then, from the Igdy wells to the Tagalek wells, the bed selects in general a westerly direction. A verst and a half below the wells of Lower Igdy, the valley of the Uzboy becomes narrower and is crossed by a rocky ridge, which has suffered little from the stream. This cataract in general resembles that of Kurtysh Its height reaches 4 sagens. From the Igdy wells to the Khalmadja wells, the character of the bed of the Uzboy is the same as for the distance from the Ak-Yaili cataracts to the Igdy wells.

In proportion as the bed of the Uzboy approaches the wells of Khalmadja, and farther on[1]), its character changes somewhat. The valley widens, and the banks of the valley and of the actual bed become even lower the sand hills, which cover the shores of the valley, become rarer and finally disappear entirely. The surrounding locality presents a tolerably even clayey saltmarsh, furrowed by rain water into ravines and crevasses. Approaching the wells of Tagalek, the bed of the Uzboy becomes shallower, from muddy deposits carried there by rain water.

From the Tagalek wells to the Dekcha springs, the bed of the Uzboy has quite a different appearance from the preceding section. It has here an immense fall, which surpasses even the first section. Beginning from the wells, the valley of the Uzboy becomes narrower, and after two versts disappears altogether. The bed here runs for about 6.5 versts through a rocky soil. Some two versts above the Burgun wells, the rocky ground ends and, to the Burgun cataracts, a distance of 14 versts, the bed of the Uzboy traverses a clayey soil.

The Burgun cataract, which begins 12 versts below the Burgun wells, in its length and height, is the most considerable cataract on the Uzboy. Its total length is about 4 versts, with a height of about 4 sagens.

In general, like the Kurtysh cataract, it presents a compact mass of rocks, obstructing the tolerably deep bed of the river. This cataract descends by several successive steep steps, of which the height of the lowest reaches 2 sagens. This step descends to a salt lake situated at its foot, 4,5 sagens deep. The rocky banks of the bed, which continue yet a verst below the cataract, and pieces of rock fallen down the slopes of the banks, testify clearly that the cataract was being destroyed by the current. The strength of the current may also be observed in the caverns formed in the rocks of the cataract itself, as a consequence of the softer parts of the rocks being washed away by the water.

The cataract, situated 5 versts above the Dekcha springs, is at present entirely covered over with mud, and only the rocky banks and the immense fall for a proportionately insignificant distance, in-

---

[1]) Six versts below the wells of Khalmadja, pieces of white marble were found in the bed of the Uzboy, carried there from the plateau of Ust-Urt.

dicate the fact, that the bed was crossed in this place by a rocky ridge similar to that above described.

At the sources of Dekcha, the mountainous character of the Uzboy ends, distinguished thus far, by a great fall and by the frequent traversing of its bed by rocky ridges. Farther on, the bed of the Uzboy turns off from the southern slope of the Ust-Urt, and right up to the Balkhan Gulf no more rocks are to be seen.

The part of the Uzboy from the Dekcha springs to its end is the most regular. The distinct characteristic of this last section of the bed is its insignificant fall, distributed pretty equally throughout its whole length, and the uniform dimensions of the bed, running along a valley, possessing a high and regular relief.

The direction of the bed from the Dekcha springs to the Tanderli wells, a distance of 170 versts, is a south-western one. It then changes it for a western direction, which it retains down to its issue into the salt-marshes depression of Bala-Ishem; the distance from the sources of Dekcha to the Tanderli wells is 170 versts. The general fall of the bottom of the Uzboy for these 193 versts is 5 sagens, the average at the end of this part being somewhat greater thant at the beginning.

The last part or section of the Uzboy abounds in a still greater quantity of spring water than the part from the Akh-Yaily wells to the Dekcha springs. The supply of water in the Uzboy is considerable because, notwithstanding a great evaporation, the dimensions of the surface of water, only slightly change. The precipitated salt is much more seldom to be seen and in a smaller quantity, in this part of the Uzboy. As a particular proof of the supply of a considerable quantity of spring water to the Uzboy, may serve the fresh water lakes met with both in the chief bed and in the old secondary beds. Of these lakes, the following deserve attention from their considerable dimensions. Lake Topiatan, situated near the wells bearing the same name and at the lower end of an old secondary bed, is one verst long. Then comes lake Tagalek, 12 versts below lake Topiatan, near the Tokhlu wells. The third lake of considerable size, is Yas-Khab, situated near the wells of the same name. Such an abundance of fresh water calls forth a rich vegetation in the valley of the Uzboy. The latter, for nearly its whole length, is overgrown with saksaul bushes, which grow so thickly, that they sometimes form an impenetrable barrier. In the neighbourhood of the lakes above mentioned poplar and tamarisk groves are to be seen. In spring, when the trees are all green, the valley of the Uzboy looks as if a river were yet flowing through it. This impression is in no slight degree helped by the presence of great quantities of wild duck, geese, boar and saygaks, with which the dense growths of cane swarm.

Three versts below where the Transcaspian railway crosses the Uzboy, the bed of the latter disapears. In this place the waters of the river used to fall into the swampy depression of Bala-Ishem, which in ancient times probably formed a continuation of the Balkhan Gulf. The bottom of the end of the Uzboy is nearly one sagen lower than the present water level. The Bala-Ishem saltmarsh now presents a shallow depression, 45 versts long and 30 versts wide, its surface being 900 square versts. It is bounded on the north by clayey sandy spurs of the Great Balkhan Mountains, on the south by offsets of the Little Balkhans, and by the Naphtha Mountain, on the south west and west by sandy hills.

The bottom consists of a porous saltmarsh, covered here and there with whole areas of crystallized gypsum.

The salt depression of Bala-Ishem communicates with the Balkhan Gulf by a river bed, which on ancient maps bears the name of Ak-Tam. This river issues from the northwestern corner of Bala-Ishem, two versts above the Kara-Durun wells.

For the first 20 versts the bed of the Ak-Tam presents now a valley with irregular sides, through the middle of which winds a narrow river-bed; it has a breadth of 20 sagens and a depth of 0,2 to 1 sagen.

The bottom of the valley is almost horizontal, rising a little above the level of the Caspian, and consists almost wholly of boggy saltmarsh; here and there are found whole patches of precipitated salt. The Ak-Tam is probably a channel, which in former times joined the Balkhan Gulf with the, depression of Bala-Ishem. The outline of the irregular banks of this bed shew very distinctly the traces of the action of sea waves. As for the little river bed to be seen in the middle of the valley, this was probably formed in much later times, namely, when the stream ceased to flow by the Uzboy and the water of the Caspian fell to its present level. From the irregular character of this bed and from the fact, that its banks are either as high as or not much higher than, the present level of the Caspian, the conclusion may be made, that this bed was formed by the tide of the Balkhan Gulf and the agitation of its waters by the wind.

Schemes for Leading the water of the Amu-Darya into the Caspian Sea: 1) By the bed of the Kunya-Darya, the Sary-Kamysh basin and by the Uzboy; 2) By the bed of the Kunya-Darya, the Shamrat Canal, the Daudan, the Canal avoiding the Sary-Kamysh Depression and by the Uzboy.

All the labours of the Expedition and the data gathered by it, completely demonstrate the possibility of letting the water of the Amu-Darya through into the Caspian Sea. No elevation of the ground between the Amu-Darya and the Caspian was found to exist. On the contrary, the country was found to fall gradually towards that sea.

The water of the Amu-Darya can be led through the steppes by the old river beds into the Caspian by two ways, by the Kunya-Darya in order to fill up the Sary-Kamysh Depression and then leading it farther by the Uzboy into the Caspian; or by constructing a canal avoiding the Sary-Kamysh Depression to the Uzboy and then by the latter into the Caspian.

Carrying into effect the first scheme, that is filling up the Sary-Kamysh basin, the water of the Amu-Darya must be sent by the old bed of the Kunya-Darya, it being necessary to take out of the Amu-Darya on an average for the year, about 70 cubic sagens per second, so that out of the basin of Sayr-Kamysh for the year, about 13 cubic sagens per second, may pass into the Uzboy. The total length of the water way between the Amu-Darya and the Caspian would be in this case about 1200 versts. The least depth of the projected water communication between the Amu-Darya and the Caspian Sea is proposed to be 5 feet, and this is based on the following combinations. The maximal draught of vessels, which can navigate the basin of the Sea of Aral, is about 4 feet, and in the Balkhan Gulf a depth of 5 feet begins only about 25 sea miles or

44 versts from the outlet of the Uzboy, so that it would be necessary to execute large works for a considerable distance so as to obtain the required depth of 5 feet. For this reason, vessels in order to navigate this waterway without impediment, must be constructed with a draught not exceeding $4^{1}/_{4}$ to $4^{1}/_{2}$ feet. The local boats called «kayuks», which at present navigate the Amu-Darya, are not more than 10 sagens long and about two sagens broad, with a draught of not more than 2 feet. In consequence of a lack of good ship-building material it is to be expected that, after the opening of the projected water way between the Amu-Darya and the Caspian, vessels like those which navigate the neighbouring basin of the Volga would be brought into use. There, the largest vessels of the «Caucasus and Mercury Company» are 40 sagens and 3 feet long and 5 sagens wide, with a draught of 4 feet 9 inches and a full cargo of 60,000 pouds on board. Therefore the minimum depth of the water way to be constructed is taken to be 5 feet, and its minimum width at the bottom, $3^{1}/_{2}$ times greater than the breadth of a vessel, that is 17.5 sagens. In constructing a steam navigation between the Amu-Darya and the Caspian Sea by means of submerging the Sary-Kamysh basin, it is proposed to bring the level of the future Lake Sary-Kamysh to the height of the present level of the Sea of Aral, 36,39 sagens above that of the Caspian Sea. It would be necessary in this case to build in the depression, situated on both sides of the Ak-Novat elevation protecting dams, to prevent the waters of the lake from penetrating into the hollow of Ikedje-Kuyu. The length of the channel on the projected lake, from the fall into it of the Kunya-Darya to the outlet of the Uzboy at Kugunek, would be 216 versts and the maximum depth of the lake would reach 45 sagens. The reason for bringing the level of the lake to a height equal to the level of the Sea of Aral is, first, that in this case on the site Kugunek at the outlet of the Uzboy, a depth necessary for the passage of steamers could be obtained without the execution of any considerable works, and then that with such a water level the surface of the future Lake Sary-Kamysh would already reach 6,000 square versts, and in case of a still higher level, that surface would also become yet greater, which in its turn would cause too great a loss of water through evaparation and filtration.

The bed of the Kunya-Darya, by which it is intented to turn

the waters of the Amu-Darya into the Sary-Kamysh basin, in its present condition thoroughly answers, for nearly its whole length the requirements of the bed of a navigable river. Only the first 39 versts from where it leaves the Amu-Darya form an exception. Here the river bed is silted up, and will require regulation for rendering it navigable. On the remaining part of the Kunya-Darya, it is only necessary to remove the dams, which exist now across the river bed, and to build a so called blind dam, to prevent the water from taking the course of the Kichkine-Darya, which forms a bifurcation of the Kunya-Darya at its outfall into the Sary-Kamysh.

For a free passage of water to the required amount out of the projected Lake Sary-Kamysh into the Uzboy, it would have to be regulated as it leaves the lake, for which purpose it would be necessary to dig four ditches, 5.65 versts long each, at the elevated parts of the Siradja and at the lower parts of this plain to build protecting dams, to prevent the formation of floods and increase of evaporation. As to the Uzboy itself, it would be necessary to deepen it at Kugunek in the above mentioned plain, for a distance of 34 versts, and to construct in its bed, near the ruins of Talai-Khan-Ata, a dam with a single chambered sluice, so as to weaken the fall of the level above.

The construction of the water route from Talai-Khan-Ata to the Caspian would be the same, whichever of the two schemes be carried out, that of filling up the Sary-Kamysh basin or that of constructing a canal avoiding the latter.

The total cost of leading the water of the Amu-Darya into the Caspian by means of filling up the Sary-Kamysh basin is calculated as follows:

1) The works on the Kunya-Darya.	5.300,000
2) » » for regulating the outlet of the Uzboy from the Sary-Kamysh basin.	3.500,000
3) Works for regulating the Uzboy from Talai-Khan-Ata to the Caspian	6.200,000
Total . . .	15.000,000

The route could be opened for steam navigation in 15 to 17 years, as the Sary-Kamysh basin could be filled with water only in this time, by letting through from the Amu-Darya into the Kunya-Darya 70 cubic sagens of water per second on an average for the year.

The second scheme for constructing a water way between the Amu-Darya and the Caspian consists in directing the waters of the Amu-Darya into the bed of the Daudan, avoiding the Sary-Kamysh Depression by a Canal about 78 versts in length, out of which the water would flow by the valley of Siradja and the Uzboy into the Caspian Sea. In this case, the water of the Amu-Darya may be directed into the bed of the Daudan within the limits of the Khivan-Khanate as far as the point, where the canal Shamrat crosses the bed of the Daudan, by the three following lines: 1) By the canal Shakh-Abat, in which case 29 cubic sagens of water per second on an average for the year might be taken from the Amu-Darya into the Daudan to flow by it and further on by the Uzboy into the Caspian, and 6 cubic sagens for the irrigation of the fields along the Shakh-Abat canal, and for the other needs of the local population; 2) by the old bed of the Kunya-Darya and the Shamrat canal, where it crosses the bed of the Daudan, when it would be necessary to take from the Amu-Darya about 24 cubic sagens for the chief principal object, and 6 cubic sagens for irrigation; and 3) by the system of the Lauzan canal, a part of the bed of the Kunya-Darya and the Shamrat canal, also up to where it crosses the bed of the Daudan, which would require about 22 cubic sagens of water per second. The lengths of the navigations in the three directions, from the Amu-Darya to the Caspian Sea, amount to respectively 1068, 1074 and 956 versts. Thus, the last direction is shorter than the two former by 113 and 119 versts.

The expense of organizing steam communication by all three routes is almost the same. By the Shakh-Abat canal and the Daudan, it is estimated at 3.700,000 roubles; by the Kunya-Darya and Shamrat, about 3.900,000 roubles; and by the Lauzan and Shamrat system, 3.200,000 roubles.

Thus, from the point of view of cost, none of the three directions has a decisive advantage over any other. Other combinations, then, must serve as a guide in the choice of the best direction. It is, accordingly, necessary to take into account, which of them is best suited for the development of agriculture, the least expensive to keep in repair in the future and, finally, the least likely to be broken through and to incur other damage during high water. Comparing the mutual merits and defects of these three directions, the conclusion is arrived at that the direction by the Shakh-Abat canal and the upper part of the bed of the Daudan is less suitable than the other two, and this for the following reason. The bed of the Daudan from its issue from the Shakh-Abat to where it crosses the Shamrat canal, for nearly the whole extent traverses an uncultivated country, and lies between sandy banks which, while their strengthening would call for considerable expenditure, would yet not be secured from obstruction and injury. The Kunya-Darya, on the other hand, and the Lauzan and Shamrat canals traverse an inhabited and cultivated country. Besides this, to the west of the Shamrat canal in the direction of the lower part of the Kunya-Darya, there lie lands capable of cultivation, lands which once formed the richest and most flourishing district of Khovarezm, but now turned into a desert. In the future, together with the development of agriculture in the Khanate, it is in the highest degree desirable to restore these lands to tillage by directing to them, through the lower part of the Kunya-Darya, the necessary quantity of water for their irrigation. As far as concerns the question of the comparative suitability of the directions as far as the Shamrat canal, by the old bed of the Kunya-Darya, or the system of the Lauzan canal, for the passage of the water of the Amu-Darya into the Caspian, avoiding the Depression of Sary-Kamysh, the route by the latter direction, notwithstanding that, is shorter and its construction less costly than that by the former, is yet not to be prefered, because the Lauzan canal system traverses a low lying region at present subject to inundation and having a general inclination westwards, and consequently presenting a possibility for the water to break through in the direction of the Aibugir Gulf, which frequently happened in former times. Besides this, the very situation of the canal, which at its exit from the Amu-Darya, represents, as it were, a prolongation of the river itself, renders the directory of a great quantity of water into it dangerous. It might happen that the whole mass of water of the Amu-Darya might break through into it.

On the basis of all that has been stated above, the best and the most reasonable way ofestablishing water communication between the Amu-Darya and the Caspian is that which, avoids the Sary-Kamysh Depression by directing the waters of the Amu-Darya by the old bed of the Kunya-Darya and the Shamrat Canal into the bed of the Daudan and thence by this bed, and a Canal avoiding the Sary-Kamysh Depression, into the bed of the Uzboy. But the defall of the country in this direction is not uniformly distributed, so that, when the schema was drawn up, different bases were taken into consideration for different sections in dependence upon their inclination. From the exit of the Kunya-Darya from the Amu-Darya to the end of the Daudan, i. e. to the beginning of the Canal avoiding the Sary-Kamysh, a distance of 315 versts, the total fall is about 19 sagens, which is equivalent to 0.06 sagen per verst, or an incline of 0.000,121, approaching very closely the incline of the Amu-Darya, and the irrigating canals of the Khivan Khanate. Accordingly, for this distance, the projected navigation is an open one. For the first 35 versts from its exit from the Amu-Darya, the old bed of the Kunya-Darya is nearly silted up with sand, so that for this distance it would be necessary to dig an artificial bed. For the next 127 versts, as far as the Shamrat Canal, the bed of the Kunya-Darya in its natural form is of sufficient size, so, that over this extent it would not be necessary to execute any works in it with the exception of the removal of the Kisil-Takyr dam. Somewhat below the exit of the Shamrat Canal, it is intended to construct a regulating dam in the bed of the Kunya-Darya, its object being to regulate the supply of water to the artificial bed, to be constructed in the direction of the Shamrat Canal, by letting through the superfluous water into the lower part of the Kunya-Darya, and thence into the Sary-Kamysh Depression. Thus the water level in the artificial bed will permanently remain nearly at the same height and the velocity of the current will be constant and not exceed the point, beyond which the banks would be apt to be washed away. Below the projected dam, along the lower course of the Kunya-Darya, as has been mentioned above, lies a very fertile country which it is intended to irrigate. For this purpose it is proposed to take out of the Amu-Darya, a greater quantity of water than will be needed for the establishment of navigation to the Caspian Sea, and to use the whole surplus in irrigation. It is intended to take advantage of the possibility of making use of the now existing canal in the direction of the Shamrat up to the end of the Daudan, only enlarging and regulating it in certain places.

For the first five versts below the Shamrat Canal, the bed of the Daudan is silted up so that there is work to be done in digging an arifical canal. For the whole remaining distance the Daudan has a more or less regular bed, in which there are no obstacles to prevent the passage of the necessary quantity of water. But the most important works, it is expected, will be on the canal avoiding the Sary-Kamysh Depression, from the end of the Daudan to the entrance into the Uzboy of the said canal at the ruins of Talai-Khan-Ata. The total fall for this length of 194 versts is 1.19 sagens or 0.006 sagen per verst, which is equivalent to an incline of 0.000012. This is insufficient to give the current in the canal the velocity needed to prevent the deposit of water-borne sediment, or the overgrowth of the canal with grass and cane. In consequence of this, it is intended to construct the canal open, so that the sections with a normal velocity of current, not exceeding that at which the ground is likely to be washed away and sufficient to prevent the formation of desposits, should alternate with levels with an insignificant fall, where these sediments are collect and be removed by dredging. The water thus freed from deposit, will flow on farther in a pure state.

From the exit of the Uzboy from the Sary-Kamish basin to the Dekcha springs, a distance of about 295 versts, the total fall is 29.88 sagens, or 0.10 sagen per verst, which constitute an incline of 0.0002. Such being the great incline for this distance, it is intended to let the waters flow partly by an open bed, regulating it at certain points, partly to construct sluices on the sections possessing a more considerable incline, by building dams across the present bed or taking advantage of the existing cataracts. Thus, there are to be in the Uzboy 7 dams and 3 falls at the cataracts of Kurtysh, Igdy and Burgun with 10 sluices in them, out of which 9 will be single and one double.

On the remaining part of the Uzboy, from the Dekcha springs to its fall into the Balkhan Gulf, a distance of 270 versts, the total fall is 6.68 sagens or 0.02 sagen per verst, which is equivalent to an incline of 0.00005. For this whole distance, it is intended to leave the bed open, at the same time to cheapen the regulation of the fall of the Uzboy into the Balkhan Gulf, the following measures being taken: 1) A canal must be constructed making a detour on the salten depression of Bala-Ishem. 2) At the fall of the Uzboy into the Balkhan Gulf, a blind dam is to be built, so as to form on the part of the Uzboy between the depression of Bala-Ishem and the Balkhan Gulf, a distance of 26 versts, a summit level, the water in which would be one sagen higher than that in the Balkhan Gulf. 3) Along the southern coast of the Balkhan Gulf is to be dug a circuit canal wide enough for the passage of one steamer, with sidings every ten versts, where steamers going in different directions may pass each other. To prevent the obstruction of the outlet of this canal into the Balkhan Gulf by the deposit of sand, it would be neccessary to construct protecting dams at its end, which could not involve any great expense, from the absence of any great agitation in the Balkhan Gulf. It may at the same time be remarked, that the minimum depth of the whole navigation to be constructed ought to be 5 feet, which would allow vessels having a draught of $4^1/_4$ to $4^1/_2$ feet, and answers, as we have already seen, to the conditions of navigation on the Amu-Darya, the Balkhan Gulf and the lover part of the Volga.

The cost of executing all these works for the establishment of a steam route between the Amu-Darya and the Caspian, by the old bed of the Kunya-Darya, the Shamrat Canal, the bed of the Daudan, the Canal making the detour of the Sary-Kamysh Depression, and by the Uzboy, in all 1074 versts in length, is estimated at about 27.000,000 roubles, i. e. about 12.000,000 roubles more than for the establishment of the same route by submerging the Sary-Kamysh basin. Nevertheless, notwithstanding the less expense of filling up the Sary-Kamysh basin with water, in comparison with the construction of the canal making the detour, it is necessary to choose the latter method, and this based on the following considerations. After filling up with water the Sary-Kamysh basin, a deep lake will be formed open to winds, with a channel of more than 200 versts in length, which shallow drought steamers will have to navigate in tow. This will not present the necessary safety, and consequently will be an obstacle to the development of trade traffic by this route. Besides this, the large surface of the lake would cause a useless expenditure of a great quantity of water in evaporation, not to mention that a considerable space of time, viz. 15 to 17 years would be needed for the establishment of this route. There is no doubt, that to avoid the inconveniences of navigation of vessels with of light-draught on the projected lake, it would be absolutely necessary at a future date to construct a canal along its shore, between the outlet of the Kunya-Darya and that of the Uzboy, a canal of not less than 160 versts in length, as was done for Lake Ladoga, and is now being done for the Lake Matko. Consequently it is possible, that the establishment of the communication by means of submerging the Sary-Kamysh depression would in such a case in the future, together with the previous original expenditure on this entreprise, cost more than 27.000,000 roubles, i. e. more than is needed now for the establishment of steam communication between the Amu-Darya and the Caspian by means of a canal making the detour of the Sary-Kamysh Depression.

CHAPTER V.

The importance of the Amu-Darya-Caspian water way.

THE extreme unsatisfactoriness of the caravan routes over the deserts of Central Asia, and the urgent necessity of joining that remote country with Central Russia by improved communications induced our Government, in 1879 and 1880, to take two decisive measures:

1) The first was the organisation of an Expedition for the exploration of the Old Bed of the river Amu-Darya between the Sea of Aral and the Caspian, in order to finally solve the question of the possibility of letting the waters of this river through into the Caspian Sea.

2) The second measure was the inauguration of the construction of the Transcaspian railway from the Caspian Sea to Fort Kizil-Arvat.

At the present moment the Transcaspian railway is not only built as far as Kizil-Arvat, but has even crossed the Amu-Darya, traversed the Khanate of Bokhara and reached Samarkand. As for the passage of the waters of the Amu-Darya into the Caspian, after five years' explorations and thorough eximination, the Expedition has decided this question in the affirmative. But nevertheless this immensely important undertaking remains to this day unrealized.

At the present moment, in view of the existence of the Transcaspian railway in Central Asia from the Caspian Sea to Samarkand, it is necessary to elucidate the question, whether it is necessary to construct the Amu-Darya-Caspian water way. This problem is easily solved by a minute examination of the mutual relation between this railway and the said water way, in the same manner as the general question of communications by rail and by water.

The question of the utility of water ways along side of railways and of the expediency of expenditure upon the construction of the former is not a new one. There was a time, when it was raised in every country of Europe, and the time is not long passed, when an unfavourable opinion upon water communication found many adherents. But as soon as the epoch of the railway fever was over, water ways again recovered the lost importance and at the present day, nothwithstanding the immense development of the railway system, the states of Europe and the United States of North America are straining every effort to extend their internal water ways, devoiting very large sums to this object.

FRANCE, which has spent more than any other European country on the development of its internal water ways, in 1883, after the completion of the chief railways, proceeded to the construction of the Marne-Rhine Canal. It extends for a considerable distance parallel to the line from Paris to Strassburg, and called forth a remarkable development of the mining, salt and iron industries in the adjacent localities. «Minerals, which had lain for many ages in the earth», remarks the engineer Picard, in his well known work upon the railways of France (Traité des chemins de fer) «wakened from» «long sleep, and issued from the bowels of the earth. Works sprung,» «as it were, out of the ground, and are now crowding against each» «other, situated as they are between the canal, which on the one» «hand brings them raw materials, and the railway, which carries» «off their manufactured products. The railway alone could scarcely» «have called forth such a wonderful development of the country.» «We see there, as in many other parts of France, a radical change» «of the face of the earth».

The year 1879 is an important epoch in the development of water ways in France. It was then that the programme of their farther improvement was elaborated. Into it entered both the amelioration of about 4,000 kilometres of rivers and 3,600 kilometres of already existing canals, and the construction of 1,800 kilometres of new canals. In the execution of this programma attention was chiefly directed to the improvement and reconstruction of the already existing navigations, and this part of the programme may be considered now as nearly accomplished, at least in the north and west of France. Although the construction of new canals, at that time, was not pushed on with the same energy, as now, nevertheless two canals, one between Havre and Tancarville at the outlet of the Seine, and the other between the Aisne and the Oise have been completed and opened to navigation some years. Two other canals, between the Marne and the Saône and between the Doubs and the Saône are still in process of construction.

The regulating works on the Garonne, executed by the engineer Farge, have produced excellent results.

For the regulation of the Rhône, from its confluence with the Saône to its fall into the Mediterranean, considerable sums have been spent.

Upon sluicing and regulating the Seine between Paris and Rouen, works terminated only in the latter half of the eighties, during the 50 years they continued, more than 100.000,000 francs were expended. But even this considerable outlay proved in practice to be quite expedient, calling forth an enormous development of navigation on this river. It was only after these works had been executed, that it became possible to convey goods from Paris to London by water alone. Fresh supplementaty works on the lower Seine are now projected, for the purpose of obtaining such a depth in this river as to enable sea going vessels to reach Paris, which will thus become a port.

We see, then, that notwithstanding the construction of railways, which have covered the whole of France with a dense network of them, the water communications of this country have been acquiring lately a greater and greater significance in the economical life of the country.

In Germany, on the regulation of the Rhine from 1831 to 1866 inclusive, all the states along its banks together have expended 114.500,000 marks. From the 1867 to 1887, the yearly outlay for the same purpose was on an average about 5.500,000 marks. Thus the total expenditare upon the execution of works for the regulation of the Rhine is about 245.000,000 marks. At the same time, during the ten years from 1875 to 1885, the traffic of goods on the Rhine nearly doubled.

On the Elbe up to 1842, no works of any importance were carried out for the improvement of its navigation, and after this till 1858, the credit assigned for this purpose was but small. But from 1859, more considerable sums began to be devoted to the regulation of the Elbe, which growing gradually, attained lately an average anomal sum of about 2.250,000 marks. Altogether, up to the present, about 58.000,000 marks have been spent upon the regulation of the channel of the Elbe. At the same time, the traffic of goods on this river during the last 15 years more then trebled. In 1875, it was equal to 435.000,000 ton-kilometres a year, and it at present reaches over 1.300,000,000 ton-kilometres. Upon the regulation of the Prussian part of the Niemen (Memel) a total sum of 10.000,000 marks has been spent, upon the Prussian part of the Vistula about 50.000,000 marks, and upon the Oder about 43.000,000 marks. The traffic of goods during this time has increased considerably.

Notwithstanding the fact, that numerous lines traverse Germany from west to east, and that the basin of the Elbe was already before joined with that of the Oder by the Friedrich-Wilhelm and Finov canals, a third canal, that of the Spree has been constructed lately, (1887 — 1890), between these basins, upon which about 13.000,000 marks have been expended.

In the eighties works were carried out for sluicing the Main from Frankfurt to the mouth of this river. Their cost was about 5.500,000 marks. During the first navigation season after the completion of the works, the goods traffic increased by nearly two and a half times, and the conveyance of goods against the stream, increased more than twenty times.

Besides what has been said above, in the immediate future, it is proposed to establish unbroken communication from the Rhine to the Elbe, the Central Germany Canal, and a canal between the Rhine and the mouth of the Ems, the scheme for which is already sanctioned. The cost of the first is estimated at about 100.000,000 marks, and of the second 60.000,000 marks. All this has been done besides the construction of the canal, between the German and Baltic seas, which is specially destined for the use of sea going vessels, the construction of which was begun in 1888.

In Austria, the works for the regulation of the Danube at Vienna carried out from 1869 to 1875 and on the Theiss at Buda-Pest, invite attention. The improvement of the conditions of navigation on the Danube at the «Iron Gates» was undertaken in 1890, and the total expense of these works is estimated at a sum of about 11.000,000 guldens.

In the delta of the Danube, from the year 1856, considerable works have been executed under the direction of a special European-Danubian Commission. The Sulina arm was chosen for regulation, and up to the present time, more than 40.000,000 francs have been laid out upon these works.

ENGLAND, the land of coal and iron, the birthplace of railways, is badly equipped from natural conditions for the construction of water ways of communication. Its rivers, with the exeption of their mouths, in consequence of their shallowness in a natural state, are unfit for any considerable navigation. But nevertheless, latterly, already after the land has been cut up in all directions by railways, there is to be noticed a very manifest movement in favour of water ways of communication. In 1875 great works were undertaken for the regulation of the Weaver, and for uniting it with the Trent-Mersey Canal.

The attention paid to water communication recently renewed, is especially evidenced by the works undertaken in 1888 by a private Company for the construction of a ship canal between Manchester and the mouth of the Mersey, near Liverpool. It is to be noted that these two centres of industriel and trading England constitute a perfect railway Kingdom. Between these points was built one of the first railways in the world. Here took place the famous competition of engines in 1829, at which Stephenson's «Rocket» won. These two cities are at present joined by five railways, some of them possess four lines of rails and nevertheless, a water way is being constructed—side by side with them. Such a victory of the water over the raylway is very significant.

The Manchester Ship Canal is destined for the navigation of sea going as well as river craft. Its cost is estimated at about 10.000,000 pounds sterling. In the construction of this canal, we see a striking example of the establishment of a new water way along side the grandest railway line in the world, relative to its transporting capacity, and this, without any competition and with the deepest conviction of the financial success of the undertaking.

America, possessing like our own country vast natural water ways has not remained behind Old Europe in its efforts to devlop them.

In 1875 considerable works were undertaken for the improvement of the mouth of the Mississippi, which were terminated with perfect success in 1879 and cost 5.250,000 dollars. Afterwards the government of the United-States turned its attention to the systematic improvement of the whole navigable part of this immense river. Preliminary investigations, unanimously acknowledged to be necessary by Congress and which subsequently served as the basis for these works, were commenced already in the fifties, by a special commission, at the head of which was Captain Humphreys and Lieutenant Abbot. After investigations lasting ten years, which often cost tens of thousands of dollars a year, Humphrey's commission gathered very valuable and circumstanted materials, which in completeness and thoroughness surpassed all investigations of navigable rivers executed up to that time. In 1879 a new commission was organized to draw up a scheme for the regulation of the Mississippi. After supplementary enquiries, which lasted about five years and cost about 650,000 dollars, a general project for the improvement of the river was drafted, estimated to cost 39.500,000 dollars, and the execution of the works was proceeded with at once. At the same time regulation works are being carried out on the Missouri, estimated at about 8.500,000 dollars.

In Russia, after the construction of the chief railway lines, it became gradually manifest that they cannot replace water ways in all respects, and that together with the building of railway lines it is necessary to look to the further improvement and the development of the water ways. First of all, in this connexion, attention was turned to the Maria Water way, the importance of which, notwithstanding the little satisfactory condition of the way itself, and the building of competing railways, was increasing from year to year.

In the year 1861 the construction of the new Emperor Alexander II's Canal, was commenced, making a detour of Lake Ladoga, side by side with the sluiced canal of the Emperor Peter the Great, which formed the most difficult spot of the whole line. These works were terminated in 1866. At the same time, the sluices were lengthened, and in the rapids of the Sheksna a chain tug was established. Next, along side the Sias and Svir Canals, also with the object of avoiding Lake Ladoga, to the east of the mouth of the Volkhov, the Canals of the Emperor Alexander II and the Empress Maria Feodorovna were built in the years 1878 to 1883.

From 1882 to 1886, a considerable improvement was effected in the system of the watershed section, by the digging of the New Maria Canal. All these works for the improvement of the Maria water system cost in all about 14.000,000 roubles. Nevertheless, the continually increasing goods traffic demanded a still farther improvement of the system in order to increase its capacity for traffic. This object is to be attained by means of the thorough reconstruction of the Maria water route, at present in course of realisation. The scheme of this reconstruction was sanctioned in the summer of 1890, and its execution was commenced at the end of the same year. The total cost of the works was estimated at 12.500,000 roubles, and five years were appointed for their accomplishment. The reconstruction of the Maria system is the first example we have in Russia of the bringing of a water way into a state completely satisfying the modern requirements of navigation, considerable altered since the origin and development of railways.

In 1883 the construction of the Jenisei-Ob water route was commenced, designed to form a connecting link in the continuous water communication from the foot of the Ural to Baikal.

When the necessity for the improvement of the water ways become evident, our Government undertook a whole series of measures with the object of elucidating the needs of this form of communication and the improvement of their condition. The gradual investigation of our whole system of internal waters was undertaken in 1875 by the Navigation Inventary Commission and the inventary parties existing under its direction. At the same time various regulation works were undertaken upon our rivers.

Since the end of the seventies, an improvement of the most difficult points on the river Pripiat has been carried out, and the regulation of the Dniepr at Kiev has been taken up, upon which about 1.500,000 roubles have been spent.

The regulative works on the Dniestr present a remarkable example from the results attained. The most difficult rapids were begun to be improved in 1884 and up to 1890 only 765,000 roubles had been assigned for these works. Notwithstanding the limited character of the sums set apart, the results attained produced a very favorable effect upon the development of the navigation. The quantity of goods conveyed from 1884 to 1889 increased nearly four fold, from 4.350,000 to 16.000,000 pouds, the navigation dues, fixed at one per cent of the value of the goods, to defray the expenses of the regulation of the river, amounting in 1884 to 28,000 roubles, reaching in 1890 the sum of 104,000 roubles. Thus these ship dues alone, levied very carelessly during some 7 or 8 years, are capable of covering all the expenses of improvement of the Dniestr.

Finally, considerable regulation works were undertaken on the Volga at Nizhni-Novgorod. At first in the beginning of the eighties a strengthening of the bank of the so-called Siberian Wharf in Nischni was carried out, and then a rugulation of the whole bed of the Volga at this town with the construction of three harbours.

These works, begun in 1890, should be completed in the current year. Taking all together, upon the improvement of the Volga near Nizhni-Novgorod, 1.750,000 roubles have heen laid out. The regulation of the Volga at the portage called «Teliachi-Brod» situated 14 versts below Nizhni-Novgorod, is a continuation of these works. This is the most dificult point for navigation on the Volga between Nizhny-Novgorod and Kazan. These works, cost of which is estimated at 2.500,000 roubles, were already commenced last year and it is proposed to complete them in 1899.

From this cursory sketch of the chief works, in connexion with water ways of communication carried out latterly, it appears that, notwithstanding the current opinion, railways cannot satisfy all the demands of every country and every state, after having been naturally led away by an enthusiasm for railways, has found itself compelled to proceed to the improvement of its water communications.

What are the reasons of such a phenomenon? What advantages do water ways possess over railways? These are questions, to the examination of which it is now necessary to proceed.

Comparing the relative merits and deficiencies of both these ways of communication, it is on the whole impossible not to perceive at the first glance, that as far as speed and regularity of conveyance are concerned, railway routes leave all other means behind and have in this respect an indisputable advantage over water ways. The steam-engine has vanguished the ancient methods of conveying goods in the rapidity of forwarding, but demands for its labour a payment considerably exceeding that, ordinarily demanded by water ways. Thus this latter method of convaying goods has retained another merit which, if not so striking as rapidity of transport, is yet not less essential to trade, namely low freight, when compared with railways. On our railways, the average rate is $^1/_{80}$ kopek per poud and verst, even a little over that, and it cannot be much reduced, as otherwise railways would be working at a loss. The lowest railway rate, namely for coal, conveyed by the truck, and over great distances is $^1/_{100}$ kopek per poud. On the other hand, water freights on an average are lower than this standard, and on ways, which are more or less well organized, where there is no risk, or delay, or transhipment at rapids, etc. river freights descend as low as $^1/_{200}$ to $^1/_{300}$ kopek and even far below this. Timber and wood conveyed in barges and rafts are estimated to cost $^1/_{1500}$ kopek per poud verst. With such freights railways of course cannot compete. To what an extent water freights are more advantageous than the railway rates, appears from the fact that notwithstanding the great development of the railway system between the Baltic and the Black Sea, a part of the goods selects the sea route from the Black Sea to St. Petersburgh round all Europe, instead of taking the shortest ways by rail.

Comparing railways and water ways, running in one and the same direction, it is evident that the conveyance of passengers, of mails, and in general of all goods belonging to the so called quick traffic, can hardly be sent in any great quantity by water, if a railway lies side by side with the latter. Water routes are also ill adapted for the conveyance of horses and cattle, and therefore this second class freight must belong to railways. Finally a third still more important kind of freights, belonging to the slow goods traffic, escapes water communication. This is manufactured goods, which are in the majority of cases forwarded in small quantities by the poud, rarely by truck loads, and require a certain speed and the certainty of their arrival at their destination at an appointed date. Although a certain share ot them falls to the water ways, yet the larger amount prefers the railways. But there remains a whole mass of goods, which must naturally belong not to the railways but to the water ways. Thus, the raw products of agriculture and mining, and timber and wood, having with great weight and size only an insignificant value, are for the most part only capable of standing very low freights. These goods are usually dispatched in great quantities and to great distances, which is convenient, nay possible, only with the cheapness of water transport. Although the railways may retain the transport of a certain portion of these bulky goods, yet they are then obliged to lower their rates to a minimum, when the amount of their clear profit becomes very small. Hence, if the railways derive some small profit by the conveyance of these freights, the competition of the water ways can deprive them of only the most trifling gain.

The mutual relation of railways and water communications was lately excellently defined by a well known French engineer M. Freycinet, in 1878 Minister of Public Works. He stated his opinion on the subject as follows: «Water ways and railways are not» «intented to exclude each other but, in the contrary, to complete» «each other. Between them there is a perfectly natural distribu-» «tion of functions. The less bulky goods, requiring speed and» «punctuality in their transport, are suited for conveyance by rail;» «they can stand better than the others the expense of such trans-» «port. To conveyance by water, are better suited goods of great» «weight and bulk and small value, which can be transported only» «when and where the freight is low. Their conveyance brings the» «railways only an immaginary profit, rather encumbering them» «than feeding them».

Examining in this light, the importance of the construction of the Amu-Darya Caspian water way for a locality, which can profit equally by this route and the Transcaspian railway, namely the Khanate of Bokhara, and the countries situated beyond it, the conclusion is unavoidable, that passengers, high speed goods, cattle, manufactured goods, sugar, and other more valuable goods, will be transported by the Transcaspian railway; while bulky goods, possessing comparatively little value, for which the choice of route depends upon the freight, notwithstanding the great distance in comparison with this railway, will certainly be sent by the

Amu-Darya Caspian water way. This appears from the following calculation.

After the development of navigation by the Amu-Darya Caspian route, the freight cannot be fixed at a higher rate than on the Maria water system between the Volga and St. Petersburgh, especially as, on the Maria water system, hardly any goods are conveyed by returning boats while, on the Amu-Darya Caspian way, a brisk traffic in both directions is to be expected. Now, let us suppose these freights to be identical, i. e. that the freight for conveying goods down the river be $^1/_{132}$ kopek per poud-verst. For the railway rate, let that be taken, which now exists for cotton transcarried by the Transcaspian railway over more considerable distances, up to 1000 versts, i. e., $^1/_{35}$ kopek per poud-verst. Then, the cost of the carriage of the above mentioned row goods, assuming the distance from Chardjui to Uzun-Ada as 998 versts, will be by rail $^1/_{35} \times 998 = 28.5$ kopeks. By the Amu-Darya Caspian water way, assuming the distance from Chardjui to the exit of the Kunya-Darya, as by the Amu-Darya, nearly 500 versts and by the Amu-Darya Caspian 1074 versts, the cost of conveyance by water will be $^1/_{132} \times (500 + 1074) = 11.9$ kopeks, i. e. 16.6 kopeks cheaper.

The freight for goods imported into Central Asia by the Amu-Darya Caspian water way, up stream, should be some what higher than down stream. Let us admit the freight to be $^1/_{116}$ kopek per poud-verst, which is nearly the difference of freight up and down the Volga. In conveyance then, of bulky goods into Central Asia will cost: $^1/_{116} \times (500 + 1074) = 13.6$ kopeks, i. e. 14.9 kopeks cheaper than by rail. That such a difference in the cost of trasport does not escape the attention of traders, is evident from the following fact. The transport of raw products between Rybinsk and St. Petersburgh by rail [1]) being some 2 or 3 kopeks on an average more than by the Maria water system, in consequence of this insignificant difference in the cost of transport, in the year 1891 37.380,000 pouds of goods were transported by water and only 15.442,000 pouds by rail, the former quantity surpassing $2^1/_2$ times the latter. Hence it is evident that raw material and bulky goods, as

[1]) The distance from Rybinsk to St. Petersburgh by the railway is 580 versts, and by the Maria water way 1074 versts.

both exported from Central Asia and imported into it via Chardjui, would be sent by water and not by rail, notwithstanding the greater distance by the former. Therefore, the construction of that water way must have a beneficent influence even on that part of Central Asia, which is already united with the Caspian by rail. But this influence will be especially great upon those parts of Central Asia, which have remained outside the railway, and are now only in a little better condition than they were, when the sole means of communication was by caravan. Such is the condition of the Khanate of Khiva, the deltas of the Amu-Darya and Syr-Darya, and finally the Aral littoral of the Kirgiz steppes. All these localities to the present moment, it may be said, are cut off from European Russia for want of convenient and improved communication. Meanwhile, the soil of the Khivan Khanate is distinguished for its fertility, and in no way yields in this respect to that of the Bokharian Khanate and of Turkestan. The Khivan cotton, renowned through out Central Asia for its fine quality, is exported out of Khiva into European Russia in but insignificant quantities. The lower reaches of the Amu-Darya present very favourable conditions for the production of rice and other cereals. Finally, the Kirgiz steppes, with their nomad population, offer every condition for an extensive export of wool, leather and other raw products of cattle raising.

The building of the Transcaspian railway was not able to affect this whole vast region, because, on the one hand, it is situated some 500 to 1000 versts from the line, and on the other, because the communication with this railway can only be effected by the Amu-Darya up stream. The current, as is known, is very rapid, and presents great difficulties for navigation against it. It is certainly entirely out of the question, that such an extensive rich country, with a population counted by millions and inexhaustible natural resources, should long remain cut off form the rest of the world. The construction of some kind or other of improved communication must be accomplished sooner or later. Its connexion in such a case by water with the Caspian will possess the most evident and indisputable advantages over the carrying of a branch line from the present Transcaspian railway to Khiva. 'Such a branch, built side by side with the Amu-Darya, will hardly be in a position to work with profit. Meanwhile, what is of most importance, the water way from Khiva to the Caspian proves to be one and a half times shorter than the route by rail via Chardjui, and therefore, apart from other advantage the water way will bring the Aral region of Central Asia incomparably nearer to the heart of Russia. The sphere of influence of the Amu-Darya Caspian water way besides the lower districts of the Amu-Darya, would include also these situated in the upper waters of that river. At the present time, these regions are in communication with the Transcaspian railway by means of the Amu-Darya at Chardjui, i. e. the point where it intersects the railway. Transshipment from the vessels into the railway trucks or vice versa is inevitable, which carries in its train besides the expenses of the actual operation, warehouse, commission and other charges. All this superfluous expenditure falls as a heavy burden on the cost of transport, not to mention the greater or less loss of time involved with the forwarding of goods, on the other hand, by water, all these additional expenses are unnecessary. The freight going one way or the other in the same ship in which it was loaded is sent on without stopping to its destination.

How much lower the freights of the Amu-Darya Caspian route must be than the railway rates for Khiva and the adjacent localities, appears from comparing the cost of transport of goods from the Khivan Khanate to the Caspian by water and via the Transcaspian railway.

Assuming as above, the freights by the Amu-Darya Caspian route at $^1/_{132}$ kopek per poud verst down stream, and at $^1/_{116}$ kopek per poud verst up stream, there results as the cost of the transport of goods, from Khiva or New-Urgench to the Caspian Sea, down the river: $^1/_{132} \times 1074 = 8.1$ kopek and, up stream $^1/_{116} \times 1074 = 9.3$ kopek. It should at the same time be observed, that the distance from the western borders of the Khanate will be some 180 versts less, and consequently the carriage there or back must be about $1^1/_2$ kopek per poud cheaper. On the other hand, the carriage of one poud up the Amu-Darya to Chardjui, and then by rail to Uzun-Ada will cost $^1/_{116} \times 500 + ^1/_{35} \times 998 = 32.8$ kopeks, i. e. four times more; while its carriage in the direction of the Caspian Khiwa will cost: $^1/_{35} \times 998 + ^1/_{132} \times 500 = 32.3$ kopeks, i. e. nearly $3^1/_2$ times more, than the corresponding rate by water.

For greater clearness, let the cost of carriage be determined for one of the most important raw materials of Central Asia, cotton, from the Khivan Khanate to the centres of cotton industry, as Moscow, Lodz and St. Petersburgh, by rail, and by water in case of the extablishment of the Amu-Darya Caspian water way and of the development of traffiic over it, on the supposition that freights in Asiatic waters are comparable with those in European. Let the railway rate in European Russia be assumed at $^1/_{45}$ kopek per poud verst, i. e. that existing for the best pressed cotton, conveyed over great distances, that for the Transcaspian railway at $^1/_{35}$ kopek i. e. the lowest existing. Let the freights for the Volga and Caspian be taken as the average of those, charged on the Volga below Rybinsk for grain cargoes up stream, i. e. $^1/_{250}$ kopek per poud verst; for the Black Sea, $^1/_{350}$ kopek per poud and verst; for the Amu-Darya Caspian way the average for the transport of grain by the Maria System, viz. $^1/_{132}$ kopek per poud and verst, down stream and somewhat above that up stream, viz. $^1/_{116}$ kopek per poud verst. Then, the cost of the carriage of one poud of cotton will be:

By water.

From Khiva, by the Amu-Darya-Caspian way, the Caspian Sea and the Volga to Nizhni-Novgorod, and thence by rail to Moscow: $^1/_{132} \times 1074 + {^1/_{250}} (850 + 2211) + {^1/_{45}} \times 410 = 29.5$ kopeks.

From Khiva by the Amu-Darya-Caspian way, the Caspian Sea, the Volga to Nizhni-Novgorod, and thence by rail to Lodz: $^1/_{132} \times 1074 + {^1/_{250}} (850 + 2211) + {^1/_{45}} \times 1757 = 52.7$ kopeks.

By rail.

From Khiva by the Amu-Darya to Chardjui, the Transcaspian rail way, the Caspian Sea, the Volga to Tsaritzyn, and by rail to Moscow: $^1/_{116} \times 500 + {^1/_{35}} \times 998 + {^1/_{250}} (850 + 561) + {^1/_{45}} \times 1010 = 60.9$ kopeks.

From Khiva by the Amu-Darya to Chardjui by the Transcaspian rail way, by the Caspian Sea to Baku, by the Transcaucasus rail way to Batum, by the Black Sea to Odessa and by rail to Lodz: $^1/_{116} \times 500 + {^1/_{35}} \times 998 + {^1/_{250}} \times 325 + {^1/_{45}} \times 840 + {^1/_{350}} \times 1000 + {^1/_{45}} \times 1246 = 83.3$ kopeks.

By water.

From Khiva by the Amu-Darya-Caspian way, tbe Caspian Sea, the Volga, the Maria System to St. Petersburgh: $\frac{1}{132} \times 1074 + \frac{1}{250} \times (850 + 2677) + \frac{1}{132} \times 1068 = 30.3$ kopeks.

By rail.

From Khiva by the Amu-Darya to Chardjui, by the Transcaspian rail way, the Caspian Sea, the Volga to Tsaritzyn and by rail to St. Petersburgh: $\frac{1}{116} \times 500 + \frac{1}{35} \times 998 + \frac{1}{250} \times (850 + 561) + \frac{1}{45} \times 1619 = 74.4$ kopeks.

Thus, the conveyance of cotton from Khiva by water proves to by cheaper than by rail; to Moscow by 31.4 kopeks, to Lodz by 30.6 kopeks, and to St. Petersburgh by 41.1 kopeks.

The Amu-Darya Caspian way will then possess such low frights for the Khivan Khanate and the Deltas of the Amu-Darya and Syr-Darya, that no railway will be able to compete with it. Under such circumstances, not merely raw materials, which naturally gravitate to water ways, but also a portion of the more valuable goods, such as manufactured articles and others, will without any doubt elect for water transport. In any case, with the realization of the Amu-Darya Caspian water way, both the export and import of raw materials will be incomparably better guaranteed for Khiva and the conterminous countries, than by rail. This circumstance is of extreme importance, as it is exactly countries richly endowed by nature, with a retarded civilization, that most of all need wares, which on bulky and of small value, such, as timber, mineral fuel and machinery, and are able on the other hand, to produce mainly raw materials, as rice, cotton and grain.

Central Asia, as is well known, is poor in woody vegetation. Its only timber, «archa» and arboreal juniper, grows high up in the mountains, and is already nearly entirely cut down in all the more or les accesible spots. The Russian and native populations of Central Asia are compelled to make use for their buldings of trees from gardens and artificially cultivated plantations, such as poplar, karagach (an elm) and tala (a willow). But these species of trees are extremely unsatisfactory from a structural point of view, on account of their porosity and tendency with which they rot so that, for buildings requiring solidity, such for example as all the railway structures and even the buildings belonging to the railway department in Chardjui, it was necessary to import timber from the Caucasus and the Volga, by rail. It is evident, •that for private persons, as in the erection of large factories, and works, the conveyance of timber by rail will be very expensive. On the other hand, by the Amu-Darya-Caspian water way, the forest wealth of the Caucasus and the basin of the Volga will find a profitable sale in treeless Turkestan.

In Cantral Asia, manufacturing industry is rapidly developing; and there is every ground to expect, that it will develop on a still greater scale in the immediate future. There have already arisen factories, works and flour mills with steam power.

If in many manufactories still use water as a motor, this is only in consequence of the extreme dearness of fuel, and the unsuitableness of the Turkestan coal, worked in the mountains of Zarevshan, and in the neighbourhood of Chimkent and Khodjent. Although petroleum is found in Turkestan, its springs are not rich and are besides situated only in the mountains in the far east in the Andizhan and Namangan districts of the Fergan territory. It is evident, that the export thence of naphtha and its residues to the Turkestan manufactories and works will cost as much as the consumption of wood fuel. The manufacturing industry of Turkestan will be placed in much more favourable conditions for its development with the supply of petroleum and its residues from Baku by sea and the Amu-Daria-Caspian water way, than it enjoyes at present. From the cheapness of Baku petroleum, it may be safely expected, that even government buildings will prefer warming with petroleum to that with the costly wood fuel, which will to a considerable extent aid in saving now devasteted groves and gardens.

For countries rich by nature and possessing all the conditions for the development of industry, the import of machinery is in the highest degree necessary and important. To this purpose, water communication is incomparably better adopted than a railway, not only in consequence of its greater cheapness, but also, because it enables machines to be conveyed ready set up.

Finally a water route will contribute in the highest degree to the export of raw material from Central Asia, so necessary to our manufactories, and obtained in such abundance from its fertile soil. Here, the first place is occupied by cotton. Before the construction of the railway, Central Asia produced yearly about 3.000,000, pouds of cotton, which was used by the local population for making fabrics of a very bad quality; only a very inconsiderable part of it was imported to Russia for the manufacturies of Moscow.

The importance of the cotton industry to Russia appears from the fact that, in 1889, more than 900 manufacturies were engaged in it, turning out goods to the value of 255.263,000 roubles, which constitutes 21.4 per cent of the whole industrial production of the country. It at the same time gave employment to 210,872 hands. Notwithstanding, howerer, the considerable development of cotton growing in Turkestan, the Russian manufactories are still in need of foreign product. From 1883 to 1888, the average yearly importation was 7.726,616 pouds. Central Asian cotton finds a competitor in the foreign, only in consequence of the difficulty of carrying it to the internal markets of Russia, the Transcaspian railway being at present the only route, by which it can be conveyed. A mass of cotton bales encumber the forwarding stations waiting for weeks together their turn to be sent off. The delay in the uninterrupted transport of cotton from the place of production, Central Asia, to the places of consumption, affects the production very unfavourably and compells the manufacturers, against their will, to turn to foreign cotton, which is imported into Russia without any delay, in the quantity needed and at the time appointed. Such a delay will no longer exist, when cotton is conveyed in by the Amu-Darya water way uninterruptedly, during the whole navigation season in Central Asia, usually broken only by a short winter, and that not every year.

After the building of the Transcaspian railway, the export of cotton began to rise considerably and reached in 1890—2.673,270 pouds, while at the same time the number of plantations increased, the quality of the cotton was improved by sowing American seeds and by the employment of more perfect machinery for cleaning.

But the development of the cotton industry to an extent, corresponding to the natural conditions of Central Asia, can and will be reached, when only the Amu-Darya Caspian water way is established.

The beneficent influence of this route is not exhausted by its significance as a cheap means of conveying goods. The advantage, which will accrue to Central Asia, Russia and the whole civilized world will far surpass that obtained from it as a powerfull lever of commerce. It is enough to direct attention to the fact, that this would also be the best means of infusing life into the vast expanses of fertile soil, which now present a sterile desert.

It is known that in Central Asia a fertile soil without water is lifeless. Without artificial irrigation, the spring vegetation, in consequence of a nearly absolute absence of rains, is already parched up by the month of May, and in summer the traveller is struck by the dreary aspect of the naked and burning steppe. On the contrary, in the neighbourhood of water, on the banks of rivers and canals, a luxuriant vegetation is developed and the soil produces crops, passing all belief. The Amu-Darya Caspian water way will serve as an excellent means to this end, because it will contain not only the water necessary for navigation but also sufficient for the irrigation of the surrounding regions. The immediate consequence then of the establishment of this communication will be the possibility of tilling the now abandoned steppes, situated to the west of the Khivan-Khanate, over an extent of some thausends square versts, or more than the present cultivated area in the Khanate.

Further more, the diversion of a part of the water of the Amu-Darya into its old bed is necessary also for preventing the yearly inundations of the Delta, which cause incalculable mischief to its population. The Delta of the Amu-Darya is all the year round under water, in the form of lakes, marshes and floods, which occupy in all about one third of its entire surface, (the delta contains about 9,000 square versts), and expends in useless evaporation and filtration from 30 cubic sagens during the low water season, to 100 and more cubic sagens of water per second during that of high-water, when the extent of the floods increases considerably. In consequence of these facts, the cultivated lands in the delta occupy hardly a tenth of its whole surface. After the diversion of a portion of the Amu-Darya's waters towards the Caspian Sea, the level of the water in the delta will fall, the floods will diminish, the marshes will dry up and a wide area fit for cultivation will be obtained.

To shew the importance of the return to life of such large expanses of land, it will be sufficient to remark, that then the lower parts of the Amu-Darya basin alone will produce 2.000,000 and more pouds of cotton. Add to this that, in Khiva, wheat and rice yield twenty-or thirty-fold, «djugara» or sorghum about two-hundred-fold, and that apricots, peaches, almonds, grapes and other choice fruits thrive there, and the importance of the restoration to cultivation of such a vast area, in the immediate neighbourhood of the Caspian and united with it by a water way, will become at once evident. There is, moreover, already in view a population for the new area to be irrigated. The Khivan, Akhal-Teke and other Turkomans, as also the Adaiev Kirgiz, have already begun to pass over to the agricultural state, and are suffering from the lock of water and irrigated lands. Besides this, the emigration movement in Russia, which grows with every year, and is now directed to the northern part of Asiatic Russia, will find for colonization incomparably richer and more convenient localities in lands, conquered from the desert and restored to irrigation, situated nearer to Russia and along a cheap and excellent water communication. The direction of Russian emigration southwards into Central Asia will bring us nearer to that country; the Russian colonists will form amidst the foregin Mussulman population a stronghold upon which the Government will always be able to rely in every emergency and which will for ever attach to Russia this country, so richly endowed by nature.

This circumstance alone imparts to the establishment of the Amu-Darya Caspian water way an extraordinary importance, the more that no other part of Central Asia presents the condition necessary for the realization of the idea. In Bokhara and Turkestan no increase of the area of cultivation can be expected, in consequence of the absence of suitable soil, and of the lock of water for the purposes of irrigation.

The scheme for the irrigation of the so called «Hunger Steppe», between Chinaz and Djizak, by the water of the Syr-Darya, by means of a great canal to be diverted from that river, proposed already by the late Governor-General of Turkestan, K. P. von Kaufmann, was quickly abandoned in consequence of the great outlay it would require. As for the vast, fertile and richly watered valleys of the Surkhan, Kafirnigan and Lower Vakhsh, the so called Ghizzars of Bokhara, they decidedly gravitate to the Amu-Darya Caspian water way, their only outlet being in that direction. All the chief natural products of these countries, wood from the mountains of Baba-Tag and Hodja-Mastam, salt from the salt springs and salines of Khodja-Ikan, pistachio nuts, «buzgunch» or the galls on the leaves of these trees, cotton from Denau, wool and fine goat hair from Baisun and the highland of Sebistan-Tau, all these are even now conveyed in Kayuks by the rivers Surkhan, Kafirnigan and Vakhsh to the Amu-Darya, for sale to the Turkomans. The establishment of the Amu-Darya Caspian water way can only increase this export and develop the economical well-being of these rich regions.

To enlarge the area under cultivation in the Teke Oasis and in Merv is also impossible, in consequence of the want of water. In the Akhal-Teke region, water is so valuable, that insignificant mountain brooks constitute the property of clans. Only the district to the west of the Khivan Khanate presents the necessary conditions for the increase of the surface under tillage. Vast expanses of fertile unoccupied land, a beneficent climate, a half nomadic Turkoman population, which lately began to devote itself with increasing energy to agriculture, finally the Russian peasent, who wanders in search of new land for thousands of versts from his home, all these conditions are ready to change the steppe desert into a fertile and flourishing oasis.

Thus the Amu-Darya Caspian water way presents such advantages as regards the conveyance of the principle raw products, and the enlargement of the area of cultivated lands, that no railway can be compared with it. The establishment of this communication alone can develop to the utmost the production forces of the northern region of the Central Asian Khanate, recall to life the now waterless deserts and unprofitable swamps and floods, and favour the general rise of cultivation, industry and trade.

This route will contribute at the same time to the export from Russia of timber, tools, machinery, wrought and cast iron, and other bulky goods, secure to our manufactories cheap Central Asian raw materials, and lastly create a permanent and thoroughly safe communication between Russia and Central Asia, free from all chance occurrences, a circumstance of extreme importance not only for commerce, but from a military point of view.

The Transcaspian railway cannot be considered as a perfectly safe mean of communication. Even if it works well and regularly while peace is reigning in the Turkoman and Merv steppes, and Russia is on good terms with Persia and Afghanistan, it certainly will not be so in case of a possible complication. Then, such security of communication by the railway could not be depended upon at all, without the employment of a great number of troops. There can be no doubt but that large Turkoman bands, moving on rapidly through the ill watered steppes, could not only hinder, but altogether stop the traffic, just at the most needed moment, when we should need to bring rapidly considerable masses of stores to different points.

The Amu-Darya-Caspian water way would pass far from the extreme limits of the Empire, so that the conveyance of goods by it would not be affected by any complications. The safety and permanence in the conveyance of stores and troops would render it possible to bring rapidly, cheaply and easily stores and forces with all its baggage. Being water, this route cannot be destroyed by a band of men. The capacity for the conveyance of goods would be unlimited, and would depend only upon the number of vessels, passing from the Volga into the Amu-Darya.

All that is above stated, fully explains the extraordinary importance of the Amu-Darya Caspian water way from an economical point of view; but it's commercial, and political importance would increase yet more should it be the means of opening communication between Russia, Europe and India. A glance at the map of Central Asia will shew how small the distance is between the basin of the Amu-Darya and that of the Indus. The town Kabul is about 620 versts from the Amu-Darya at Khulum, i. e. the distance from the Amu-Darya to Peshawer, the last station of the Indian railway system in the northwest, is about 850 versts. This route admits wheel communication, possesses fresh water, and traverses an inhabited country. A considerable trade now exists by it between Kabul and Bokhara. Another road goes from the fortress of Ish-Kashim, by a navigable tributary of the Amu-Darya, the Piandj, towards Chitral and Peshawer, a distance of about 350 versts. Finally, a third passes from Kala-Piandj, by the river Piandj, the passes of Barogil and Darkot, to the towns of Yasin, Gilgit and Bundj on the bank of the Indus, a distance of about 300 versts. Thus the last station of the Indian railways in the Pendjab is distant from 350 to 850 versts, from the Amu-Darya. Remembering that the commerce between India and Europe reaches a sum of 1.000,000,000 roubles yearly, the goods being carried on board of more than 3000 ships round Asia and Europe by the Suez Canal, that a considerable part of the colonial goods consumed by Russia to the amount of several tens of millions of roubles, is brought to us from India, so that these goods are obliged to travel some tens of thousands of versts from Bombay and Kurachi to Hull and thence to the Baltic ports; that many costly goods would choose an internal water way in prefernce to the dangerous navigation in the Indian Ocean and the Red Sea, and finally that many goods, before being loaded in Bombay or Kurachi, must be brought to these ports a distance of more than a thousand versts from Lahore, Peshawer and other points of the Pendjab, — it can scarcely be doubted, but that a considerable part of Indo-European commerce, instead of being directed to the ports of the Dakka peninsula and thence through Suez to Europe, would prefer the Amu-Darya and the Caspian Sea, on account of the perfect safety of a river navigation and the comparative cheapness of freights.

That this view is well founded, it is enongh to recall those numerous schemes for the establishment of communication by rail between India and Europe, which have been put forth since 1835, when for the first time investigations were made in Persia and Asiatic Turkey by Lord Chesney. Nay, even at the present time an agitation exists in England in favour of constructing a railway from the Bosphorus to Shikarpur. If the projectors of Indo-European railways count upon the attraction of a part of the Indo-European freights, why should they not take the Amu-Darya Caspian water way, the conveyance by which will always cost less, than by the railway, the construction of which from the Bosphorus to Shikarpur, for a distance of more than 3500 versts, will require as much as 200,000 francs per verst, according to the estimate made be Lord Chesney, or as much as 280.000,000 roubles for the whole distance, at the exchange of one franc = 40 kopeks. There is no doubt but that, in case of the development of communication from the frontiers of Afghanistan by the Amu-Darya, the Caspian, and the Volga to the very heart of Russia, a railway will be built between the Amu-Darya and Peshawer in consequence of the extreme importance of such an uninterrupted steam road. There would then be an Indo-Amu-Darya-Caspian route, from the Indian Ocean through Northern India, the Afghanistan, Central Asia and Russia to the Black Sea and Baltic, a route, of which a small part only would not be within the limits of Russia, and which would without doubt present more permanence and cheapness, than a continuation of the Transcaspian railway to Shikarpur, because the first direction would chiefly consist of water ways with low freights within the limits of our own dominions and traverse fertile Central Asian Khanates, while the prolonged Transcaspian Railway with its high rates, would pass through steppe deserts and thinly populated regions. Only after the construction of the Indo-Amu-Darya-Caspian route, will the productions of our industry be able to penetrate into the depths of India. If even at present in the southern ports of India and the Island of Java, about 6.000,000 pouds of foreign petroleum are consumed annually, there is reason to suppose that, if our industry had a cheap road, Baku petroleum and its products would find numerous and profitable markets in Northern India, the conveyance to which from the ports of Southern India now costs so much. There can be no doubt, but that the consumption of petroleum in Northern India will exceed the figure of 6.000,000 pouds per annum, now consumed in the Southern India and Java. Besides this, Russia consumes various imported goods to the colossal sum of about 562.000,000 roubles yearly, much of which, as cotton, indigo, silk, sago, coffee, pepper, gum, incense caoutchouc, ginger and other spices are imported in considerable quantity from India. It must be supposed that, after the establishment of the Indo-Amu-Darya-Caspian route, our industry will be in a position to receive all these goods direct at first hand and that Russia will not be obliged any more to pay foreigners freights and commission for the carriage of goods from India through the Suez Canal round all Europe. In conclusion, it may be added, that the very geographical situation of the future Indo-Amu-Darya-Caspian road shows, that relations between Russia and Europe and India can arise and become firmly established by this line, as the shortest and cheapest of all. By every other, the goods have to be conveyed by land such a distance, that their carriage will be very expensive. Consequently, no railway from India through Asia into Russia and Europe, to the shores of the Black and Mediterranean Seas will be able to compete with the Indo-Amu-Darya-Caspian route.

An examination of the chief possible lines of railways from India to Europe, and the centre of Russia, leads to the following results.

From India to Europe.		By rail versts.	By water versts.	Total.
1. Indian - Asia - Minor railway	From Shikarpur, through Persia and Asia Minor to Skutari on the Bosphorus . .	4300		4300
2. Indian - Transcaspian route	From Shikarpur, trough Herat, Krasnovodsk, Baku and Batum.	3250	325	3575
3. Indo-Amu-Darya-Caspian route	From Peshawer, to the Amu-Darya, by the Amu-Darya and the Caspian, Baku and Batum[1]). . . .	1700	2300	4300

	By rail (versts)	By water (versts)
[1]) From Peshawer to the Amu-Darya. . .	850	—
By the Amu-Darya to New-Urgench . .	—	875
» » Old bed of the Amu-Darya . . .	—	1074
» » Caspian Sea to Baku	—	325
» » Transcaspian railway from Baku to Batum	840	—
Total. . . .	1690	2274.

Or, in round numbers, 1700 versts by rail, 2300 by water.

As to communications between Russia and India, the advantages of the Amu-Darya way are here still more evident. Thus choosing as termini Moscow and St. Petersburgh, and the towns situated on the banks of the Indus, Shikarpur and Peshawer the following results are obtained.

From India to Russia.		By rail versts.	By water versts.	Total.
1. Indian-Caucasian railway	From Shikarpur, through Persia, the Caucasus, Baku and Petrovsk to Moscow.	5900		5900
	Ditto, to St. Petersburgh	6100		6 10
2. Indian - Transcaspian route.	From Shikarpur, through Herat, Krasnovodsk, the Caspian, Volga and Oka, to Moscow[1])	2400	3950	6350
	Ditto to St. Petersburgh [2]).	2400	4500	6900
3. The Indo-Amu-Darya-Caspian route .	From Peshawer to the Amu-Darya, by the Amu-Darya, the Caspian, Volga and Oka, to Moscow [3]).	850	6000	6850
	Ditto, to St. Petersburgh [4])	850	6550	7400

	By rail.	By water.
	versts.	
[1]) From Shikarpur through Herat to Krasnovodsk . . . . , ...	2400	—
By the Caspian, from Krasnovodsk to the mouth of the Volga	—	760
By the Volga, from its mouth to Nizhni-Novgorod.	—	2211
By the Oka and Moskva, from Nizhni-Novgorod to Moscow.	—	998
Total. .	2400	3969

Thus to shorten the conveyance by land by some thousand versts is only possible by the construction of the Amu-Darya-Caspian route. The peculiar importance of this fact becomes still more evident, if it be borne in mind, that even on European railways the average distance traversed by goods, is from 400 to 600 versts, above

---

Or, in round numbers, 3950 versts by water.

	By rail.	By water.
	versts.	
2) From Shirkarpur through Herat to Krasnovodsk.	2400	—
By the Caspian, from Krasnovodsk to the mouth of the Volga	—	760
By the Volga, from its mouth to Rybinsk	—	2677
By the Maria system, from Rybinsk to St Petersburgh	—	1075
Total . .	2400	4512

Or, in round numbers, 4500 versts by water.

	By rail.	By water.
	versts.	
3) From Peshawer to the Amu-Darya	850	—
By the Amu-Darya to New-Urgench . .	—	875
By the old bed of the Amu-Darya . . .	—	1074
By the Caspian to the mouth of the Volga	—	850
By the Volga, from its mouth to Nizhni-Novgorod	—	2211
By the Oka and Moskva, from Nizhni-Novgorod to Moscow	—	998
Total . .	850	6008

Or, in round numbers, 6000 versts by water.

	By rail.	By water.
	versts.	
4) From Peshawer to the Amu-Darya	850	—
By the Amu-Darya to New-Urgench . .	—	875
By the old bed of the Amu-Darya . . .	—	1074
By the Caspian to the mouth of the Volga	—	750
By the Volga, from its mouth to Rybinsk	--	2677
By the Maria system, from Rybinsk to St. Petersburgh	—	1075
Total . .	850	6551

Or, in round numbers, 6550 versts by water.

which they shew tendency to abandon the railways, while by water ways, even valueless raw products, traverse thousands of versts without any considerable increase of their price.

Even if the Transcaspian railway is subsequently continued from Merv to Quetta, the last station of the Indian railways on the frontier of Afghanistan, it cannot be a dangerous competitor of the Indo-Amu-Darya-Caspian route, because each of these roads has its own sphere of action. The Indo-Transcaspian railway will have its sphere of action in South Afghanistan, the northern part of the Dakka peninsula and Beluchistan, while that of the Indo-Amu-Darya-Caspian would be all Central Asia with Kashgar, Northern Afghanistan, Pendjab, Northern India, Lahore, Multan, Delhi etc.

Besides all that has been stated above, the Indo-Amu-Darya-Caspian route possesses yet one great advantage over all other Indo-European railways. The Indo-Transcaspian railway would run, partly along our extrem frontier and partly through Afghanistan. The Indo-Asia-Minor railway would for its whole length, and the Indo-Caucasus railway for about 2,500 versts run outside the limits of our dominions. For these reasons, these roads would not only not be very advantageous for Russia, but the maintenance of communications by them would be extremely difficult and even impossible, while the Indo-Amu-Darya route would require to be guarded only for a distance of about 400 versts, i. e., from the Hindukush to Peshawer, because it would pass for nearly its whole length within the limits of our dominions.

Thus all the interests of Russia and Central Asia demand the immediate establishment of a water communication by the Amu-Darya from the Hindukush to the Caspian. Being then joined by rail with Peshawer, it would form an Indo-Amu-Darya-Caspian route of immense importance for the intercourse of the 300.000,000 population of India with that of the 300.000,000 of Europe. This route would join the Indian Ocean with the Black Sea and Baltic, which would open to our trade new markets in India, and would renew the trade of India with Europe through Central Asia and Russia, as was the case in the earliest antiquity.

ERRATA.

Page.	Line above	Line beneath	Instead of	Must be
III	—	17	eack	each
V	7	—	Chau	Chan
VIII	—	2	by-word	nick-name
XIV	11	—	af er	after
»	—	1	fo	of
2	14	—	and marshy	but not marshy
6	—	1	bitter	salt bitter
7	19	—	Tahenia	Taheria
8	14	—	upon the site	about the site
11	8	—	great basin	wide depression
12	—	13	to 76—67 cubic	76—67=9 cubic
13	10	—	What	what
—	—	3	he can regulate	the regulation of
—	—	1	only by means	can be obtained only by means
22	—	17	XIV	XIII
23	—	17	Can lyeth	Can Azym dwells
—	—	18	The southern part	On the southern part
25	17	—	neighbourhood	neighbourhood of Urgench
—	—	15	castle	fort
—	—	13	castle	fort
—	—	9	castle	fort
30	5	—	found three dams	found three other dams
35	13	—	intothe	into the
48	11	—	flowiug	flowing
50	—	5	Nenxt	Next
51	—	5	tream	stream
62	12	—	Sary-Kamysh basin	Sary-Kamysh Depression
65	9	—	188	1881
65	—	14	The 24 of January	The 14 of February

Page.	*Line* above	beneath	*Instead of*	*Must be*
76	9	7	3 versts	13 versts
77	11	—	About 10 versts	About 8 versts
79	—	13	19 versts	14 versts
80	3	—	about 30 versts	about 38 versts
—	3	—	it	in
—	17		reaches 9 sagens	reaches from 0.4 to 0.8 sagens
85	2		ovet	over
—	—	15	About 32 versts	About 22 versts
88	6	—	the Sary-Kamysh basin	the Sary-Kamysh Depression
—	7		and the Uzboy	and by the Uzboy
—	9	—	the Sary-Kamysh basin	the Sary-Kamysh Depression
—	—	—	and the Uzboy	and by the Uzboy
—	11	—	the Sary-Kamysh basin	the Sary-Kamysh Depression
—	—	—	and the Uzboy	and by the Uzboy
—	13	—	the Sary-Kamysh basin	the Sary-Kamysh Depression
—	14		and the Uzboy	and by the Uzboy
89	5	—	and the sea of Aral	and on the north by the Sea of Aral
—	6	—	eastwards	.westwards
—	14	—	population	settled population
90	4	—	ther are more	there are no more
91		9	3 to 4 sagens	0.3 to 0.4 sagens
94	7		second ary	secondary
—	14		with floods, which	with floods overgrown with cane, wich
—	—	14	the are low	the are sloping
95	1	—	quits the Amu-Darya below the	Quits the Amu-Darya twelwe and a half versts below the
96	19	--	or Divan-Begi	from where the waters, stupped up in this bed, are directed in the Canal Kizil-Takyr or Divan-Begi

MAP
OF THE AMU-DARYA-CASPIAN WATER ROUTE
Black Sea
Kizil-Kum
Explanations

MAP
OF THE OLD BED OF THE RIVER AMU-DARYA,
OF THE KHIVAN KHANATE,
AND OF THE DELTA OF THE AMU-DARYA.
Scale: 1:2,100,000
Sea of Aral
Kizil-Kum Sands
Ust-Urt Height
Sary Kamysh Depression
Kara-Bugaz Gulf
Caspian Sea
Explanations.

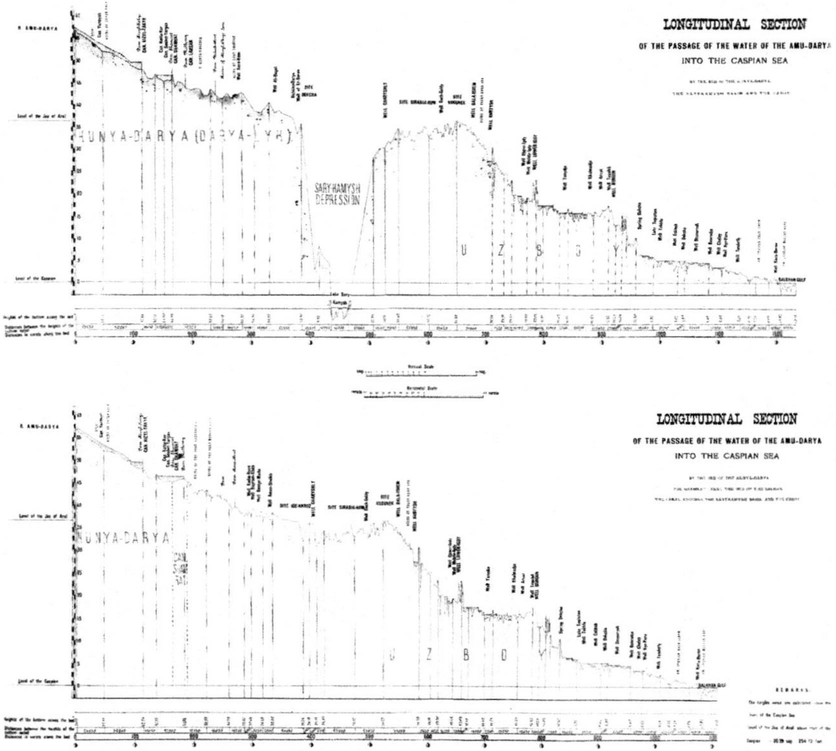
LONGITUDINAL SECTION
OF THE PASSAGE OF THE WATER OF THE AMU-DARYA
INTO THE CASPIAN SEA
SARYKAMYSH DEPRESSION
U Z B O Y
LONGITUDINAL SECTION
OF THE PASSAGE OF THE WATER OF THE AMU-DARYA
INTO THE CASPIAN SEA
U Z B O Y
REMARKS

Printed in Dunstable, United Kingdom

91230737R00100